Springer Aerospace Technology

The *Springer Aerospace Technology* series is devoted to the technology of aircraft and spacecraft including design, construction, control and the science. The books present the fundamentals and applications in all fields related to aerospace engineering. The topics include aircraft, missiles, space vehicles, aircraft engines, propulsion units and related subjects.

More information about this series at http://www.springer.com/series/8613

Kuklev E.A. · Shapkin V.S. ·
Filippov V.L. · Shatrakov Y.G.

Aviation System Risks and Safety

Springer

Kuklev E.A.
Saint-Petersburg, Russia

Shapkin V.S.
Moscow, Russia

Filippov V.L.
Moscow, Russia

Shatrakov Y.G.
Saint-Petersburg, Russia

ISSN 1869-1730 ISSN 1869-1749 (electronic)
Springer Aerospace Technology
ISBN 978-981-13-8124-9 ISBN 978-981-13-8122-5 (eBook)
https://doi.org/10.1007/978-981-13-8122-5

This Springer imprint is published by the registered company Springer Nature Singapore Pte Ltd.
The registered company address is: 152 Beach Road, #21-01/04 Gateway East, Singapore 189721, Singapore

Preface

This monograph is one of the first publications attempting to find a solution to the "rare events problem" without using directly any methods of the classical reliability theory and probability theory. As a basis, it is proposed to adopt a methodology for calculating risks as a "hazard measure" (Institute of Control Sciences of RAS) and an expert approach to defining system safety indicators through the "hazard condition" within methods using Fuzzy Sets (Fuzzy Sets [1, 2]). In this regard, the book proposes a new scientific doctrine called "Reliability, Risks, Safety" (RRS) by the authors.

The authors are aware that the approaches described in the book should be considered as the first steps in the chosen direction, and work of this kind can be continued, especially since the development of "risk models" in various fields of CA activity has demonstrated very significant results.

The main object of the RRS study are random (rare) events occurring with a "near-zero" probability and having a negative result (in adverse impact) for the systems operated. Such events are classified in a broad sense as "catastrophes", the number of which is small, but the consequences from them are very significant.

The problem of "rare events" is declared by the International Civil Aviation Organization (ICAO) as one of the most important in the domain of issues related to ensuring the flight safety.

The models of catastrophic phenomena proposed in this paper have nothing in common, except the name, with the well-known theories of catastrophes studied on the basis of the V. Arnold's concept [3], where the concept of a catastrophe is a characteristic of the bifurcations of dynamic processes arising under special conditions of the connections between elements in structures of the given (known) systems under study (with some states of homeostatic equilibrium, according to G. Malinetskiy [4]).

In catastrophe models by V. Arnold, the number of variables (and processes) that determine the surface of homeostatic equilibrium parameters does not exceed 2–5, which does not reflect the properties of actual (real-world) systems. Therefore, different approaches are needed for solving the issues of ensuring system safety when rare events (with damage) occur in systems with a multitude of high-dimensional parameters and random external influences that determine the significance of the risks of serious negative consequences, including those in technical dual-purpose complexes.

According to ICAO statistics [5], the number of catastrophes and accidents involving Airbus aircraft with various flight experience is 1–2 for 10–15 years of operation (for various aircraft classes) with a total "operation" time of more than 10 million take-offs and landings. Therefore, for example, such a concept of "risk" as "the probability of a negative event" is already inappropriate regarding the real life.

Hypotheses about the measurability of random events in the sense of the axiomatics of probability spaces (according to A. Kolmogorov) cannot be applied in the strict sense to the solution of "rare events" problems. The values of the physical parameters that determine the occurrence of catastrophic phenomena are very small and lie in the region of "heavy tails" of the probability density function (pdf). Probability distribution function (prdf) exact analytic expressions with "heavy tails" have not been found, and "rare" events of this type are immeasurable (according to A. Kolmogorov). Therefore, the hypothesis of "fuzzy measurability" of rare events (with a "near-zero" probability) has to be considered the main working postulate in the RRS doctrine.

Thus, it turns out that the RRS is applicable for the study of highly reliable systems where the high quality required by consumers is guaranteed by high-standard indicators like "reliability probability" with values of about "one". But the "consumer community" requires (*legislatively*) the provision of precisely "high reliability of systems". First of all, this requirement automatically leads to a decrease in the probability of undesirable functional failures in systems, which is exactly what the community needs. But this gives rise to another problem, the occurrence of hazardous (risk) events in the form of "rare events". From this follows the need to study other properties of technical systems as well (except quality), namely the "probabilities" of loss of functional properties by systems, even in very rare cases, but with great "damages". The book shows that it is most convenient and effective to solve the problem with the help of the above-mentioned "Fuzzy Sets" approach. But the number of investigations with this new approach is still very small.

The *background* for the creation of the RRS doctrine and its tools in the form of the system safety theory (SST) are set out below.

Development and improvement of safety systems for various complex technical systems, and not only aviation ones, should be considered as the main goal for transport systems, as well as, for example, nuclear power plants and complexes for the coming decades.

The common difficulty here is the "rare events problem" and the calculation of risks with insufficient statistics.

The priority of creating highly reliable systems was stated in FAA documents [6, 7], where A. Younossi pointed to the relationship between QMS and SMS. The same idea was taken as the basis for the RRS doctrine that, in this manner, arose as a response to the "challenge" of the world aviation community: "to transit to the provision and monitoring of ATS safety based on the calculation of risks and the theoretical SST methods" set out, inter alia, in this book.

The NASA materials [8], which the authors of the book managed to get acquainted with, are of great importance for substantiating the RRS positions. The fact is that only sufficient NASA resources allowed for large-scale stochastic imitation modeling based on the properties analysis of complex systems, such as pdf and Prdf. These results presented partially in Chap. 3 of this book, made it possible to confirm the validity of the hypotheses regarding the essence of the "rare events" problem. For this, we had to refuse, as far as possible, from probabilistic indicators and adopt the "Fuzzy Sets" approach.

In this regard, it can be argued that the "rare events" problem in civil aviation does not exist anymore, or more precisely, this problem persists only in the PSA.

Similarly, it can be stated that the human factors (HF) problem in its traditional interpretation [9] can be solved more effectively if the idea of separate study of the objective technical and economic conditions for the occurrence of "catastrophes" and the study of the possibilities (or motivation) for operators' properties is considered to be true.

Here, it is worth noting that there is a certain similarity in solving issues of the "rare events" problem in transport (e.g., in civil aviation and JSC Russian Railways (JSC RR)): a transition to fuzzy indicators of normalized frequencies of occurrence of random (hazardous) events in the methodology of calculating risks. But in the practice of civil aviation and JSC RR, it is stated (there is even a "risk analysis step" in [10] that *the probabilities of certain events can be assigned*, up to values of 10^{-7}–10^{-12}). GOST R 51901–2002 standard (*Reliability Management*) states that events with a *probability* of 10^{-6}*are already next to almost impossible*. According to the TFS provisions of the TSB (according to the RRS concept), *only acceptable levels of risk can be assigned*, for example, through the same frequencies (above). The values of frequencies can be taken from the general statistics (*in various probability spaces, probably*), as was shown for the first time in the work of M. Kumamoto or be assigned, but in a *fuzzy way*—through the *Fuzzy Sets*.

Probabilities must be proved and defined in a given probability space on the basis of the analysis of objective properties of the systems under consideration.

In this context, this book can be useful as the first step to find ways to solve the "rare events problem", according to the ICAO, but on the ideas of NASA.

The first edition of this monograph was published in the publishing house of FSUE State Research Institute of Civil Aviation (GosNII GA) in 2013.

Some additional important notes from the authors of the presented manuscript. In the book, the perspective scheme to expose main position of universal theory of aviation safety and security based on risk is demonstrated—approach of ICAO & NASA according to Fuzzy Sets (Berkley School). Probabilistically, positions of known PSA are denoted in the case of the situation with systems so as the "rare events". Using of the *given probable space* is not corrected here. The famous work CATS (Causal Model for Air Transport Safety) proposed conception of cause-and-effect chains based on formal logic rules. It is proposed to conserve only the matrix (1.4) from the page 13 or 15, named as H or {X} and after this to find the *EQUATIONS* of *catastrophes* possible in systems under danger factors. *Thus Boolean distributive lattice is the foundation of probabilistic conception of* *reliability theory that allows using of Kolmogorov's axiomatic by means of models of probabilistic spaces. But in case of rare events it is irrational and incorrect. Therefore the risk-oriented approach based on Fuzzy Sets and non-distributive Boolean lattice was adopted in safety system theory.*

There is last publication for the theme: E. Kuklev, V. Zhilinskiy/2018. Accident Risk Assessment for Highly Reliable Aviation Systems in Emergency Situations// Transport and Telecommunication Vol. 19, no. 1, 59–63. Transport and Telecommunication Institute, Lomonosova, Latvia https://doi.org/10.2478/ttj-2018-0006.

Riga, Latvia Kuklev E.A.
2018

References

1. Orlovskiy SA (1981) Problems of decision making with fuzzy source information. Science FM, Moscow (in Russian)
2. Rybin VV (2007) Fundamentals of the fuzzy sets theory and fuzzy logic. Study guide. STU Moscow State Aviation Institute, Moscow, p 95 (in Russian)
3. Arnold VI (1995) Catastrophe theory. Science FM, Moscow (in Russian)
4. Malinetskiy GG, Kulba VV, Kosyachenko SA, Shnirman MG et al (2000) Risk management. Risk. Sustainable development. Synergetics. Series "Cybernetics", RAS. Nauka, Moscow, 431pp (in Russian)
5. Documents of the 37th ICAO Assembly (Oct 2011, Montreal)
6. Younossy AM (2012) 10 things you should know about safety management systems (SMS). SM ICG, Washington
7. SMM (Safety Management Manual): Doc 9859_AN474—Doc FAA: 2012
8. Probabilistic Risk Assessment Procedures for NASA Managers and Practitioners—Office of Safety and Mission Assurance NASA. Washington, DC–Aug 2002 (Version 2/2).
9. Kozlov VV (2008) Safety management. OJSC "Aeroflot". Moscow (in Russian)
10. Kuklev E, Zhilinskiy V (2018) Accident Risk Assessment for Highly Reliable Aviation Systems in Emergency Situations. Transp Telecommun 19(1): 59–63. Transport and Telecommunication Institute, Lomonosova, Latvia. https://doi.org/10.2478/ttj-2018-0006.

About This Book

The paper describes general principles of assessing the operational safety of complex aviation and technical systems from the perspective of the methodology for calculating risks of occurrence of negative random rare events, such as crashes or catastrophes (or serious aviation accidents, e.g., with civil aircraft). For the first time, a general scheme for solving the problem of rare events is proposed that is based on approaches using Fuzzy Sets. Such an approach makes it possible to assess risk appearance of possible consequences and to solve quite reasonably a number of problems using the analysis of properties of random events with *nearly - zero* or "almost -zero" probability of occurrence. The paper also substantiates the necessity for developing a new doctrine for assessing the system safety within the "Reliability, Risks, Safety" paradigm, which in a number of special cases can be adopted as an alternative and complementary to the well-known traditional method of probabilistic safety analysis ("PSA"), which is the main analysis tool in the classical reliability theory for the system safety under the ICAO concept.[1]

In the book, so traditional methods such as "Poisson distribution" and "Bayes approach" are completely denoted because the one is "dubious tool" for extracting some information from "zero".

For specialists in aviation activity safety and flight safety.

Kuklev E.A., Professor, Doctor of Technical Sciences
Shapkin V.S., Professor, Doctor of Technical Sciences
Filippov V.L., General Director
Shatrakov Y.G., Professor, Doctor of Technical Sciences

[1] ICAO—International Civil Aviation Organization (Montreal—Canada).

Introduction

The global tendencies in the development of science and safety of various systems are characterized by the transition to the principles of calculating risks and studying catastrophes, in particular, in man-induced phenomena (the catastrophes at Fukushima 1, air crashes, accidents at hydroelectric power plants, mines, etc.).

The state of scientific developments in the field of *safety issues* is such that only in civil aviation (regarding the world aviation community) there are sufficiently promising results in identification of the risks of damage and substantiation of the ways to manage the operational safety of aviation and technical systems.

Not least because of this, ICAO (in the field of civil aviation) at the 37th Assembly (2010—Montreal) promulgated the Declaration on the need and importance of addressing the *rare events problem* in connection with investigations of aircraft accidents.

ICAO also proposed to *adjust* the traditional *safety stereotypes* like *if it is reliable, then it is safe*.

However, to investigate safety problems, no appropriate scientific apparatus was proposed, except for the classical reliability theory and its branch in the form of probabilistic safety analysis (PSA).

In this connection, it is shown in this book that rare events problems that cannot be properly solved with the methods of "probabilistic safety analysis" (PSA) should be considered from the standpoint of occurrence of random (uncertain) events with a *near-zero probability* based on a *new doctrine* named "Reliability, Risks, Safety" in this paper.

Here, a detailed interpretation of the mathematical model of risks is applied in the following form: *Risk is a measure of the amount of hazard* for a condition of the system when a random (uncertain) discrete event may occur that may have undesirable consequences or cause damage.

The fuzziness of subsets of the values for risk factors and safety indicators is crucial in the theory of the safety management system. All this reflects the methodology for determining the significance of risks and the application of risk analysis matrices (according to ISO, ICAO) based on the methodological provisions of the Fuzzy Sets (and subsets) theory.

The study considers the interpretation of the introduced concepts within the Fuzzy Sets concepts of M. Fujita (*Tails—far from medium*) and G. Malinetskiy (Hard Tails from Risk Theory—RAS ICS). That is why in the ISO and ICAO documents, solutions of practical and theoretical problems are recommended a priori without using the probabilities of events, but with indication of the use of possible event frequencies instead of theoretical values of probabilities that are unknown.

This is true for the analysis of single catastrophes that occur in aviation and in water transport, for example, in situations like that with the "Concordia" liner, when there is no reliable statistics (these examples as basic ones are described in this book).

The new doctrine (RRS) and the SST considered in this work provide for the preservation in a full and unchanged form of the methodology for ensuring the reliability of the systems as a whole, which makes it possible to achieve the required standard level of reliability indicators and "residual risk" with a probability not worse than 10^{-6} per given risk event to guarantee low levels of catastrophe risks of up to 1–3 accidents over 10–15 years of operation, such as for Airbus aircraft.

The basis for the RT to SST transition is the use of categories of fuzzy subsets. This means that the logical deterministic hypercube of the event truth in catastrophic situations is replaced by a vector hyperspace in fuzzy subsets.

The method proposed in the book is not opposed to the traditional PSA (probabilistic safety analysis) method, but it is considered as an alternative method for solving those problems of the rare events class that are difficult and cumbersome to solve based on the PSA procedures and apparatus.

The authors do not claim completeness of a number of issues considered in the book. The only goal was to present the available results and, if possible, to substantiate the need to develop another additional approach to assessing the level of safety of technical (aviation) systems and to managing risks in the Fuzzy Sets class on the basis of the ICAO (NASA) concept of proactive control of the system state when risk factors change.

The most significant contribution to the solution of problems in this domain was made by such well-known specialists as N. Makhutov, I. Aronov, S. Mikheev, N. Severtsev, E. Barzilovich, G. Malinetskiy, V. Kashtanov, V. Korolev, V. Bening, S. Shorgin, S. Orlovsky, A. Orlov (Bauman Moscow State Technical University), N. Sirotin, B. Zubkov, V. Rukhlinsky, S. Kabanov, S. Daletskiy, etc. But the "rare events" problem has not yet been resolved.

References

1. Volodin VV (ed) (1993) Reliability in technology. Scientific-technical, economic and legal aspects of reliability. Blagonravov Mechanical Engineering Research Institute, ISTC "Reliability of Machines"—RAS, Moscow, pp 119–123 (in Russian)

2. Ryabinin IA (1997) Reliability, survivability and safety of ship electric power systems. Kuznetsov Naval Academy, St. Petersburg (in Russian)
3. Novozhilov AB, Neymark MS, Cesarskiy LG (2003) Flight safety (Concept and technology). Mechanical Engineering, Moscow, 140pp (in Russian)
4. Aronov IZ (1998) Modern problems of safety of technical systems and risk analysis. Standards and quality, №. 3 (in Russian)

Contents

Abbreviations

$\bar{R}$	Average risk (scalar in the traditional form of RT records)
$\hat{R}$	Integral value of risk as a function of a set of elements in $\tilde{R}$ (implication into the ICAO risk matrix)
AC	Aircraft
ACS	Automated control system
ATC (ATM)	Air traffic control
CA	Civil aviation
CALS/PIS (Continuous Acquisition and Life-Cycle Support)	Concept of information support for the life cycle of products
F1, F2	Factors (characteristics) in groups (1–structural safety, 2–operational safety)
FF	Functional failures
FR	Functional reliability
FS	Flight safety
ILS	Integrated logistic support
IS	Industrial safety of 3 types in the MIC
LC	Life cycle
LCC	Life cycle concept
M&R	Maintenance and repair
MEL	Minimum equipment list
MSG	Product replacement strategies for maintenance and repair according to the technical condition control method
PF	System performance function

QMS	Quality management system
R	Risk (hazardous) event: $\tilde{R}$—risk assessment in the form of multiple elements (tuple) according to the ICAO concept
R&D	Research and development work
RT	Reliability theory
SS	System safety
SST	System safety theory

Chapter 1
Assessing the System Safety Using Reliability Theory and PSA Methods

The materials in this chapter do not contain any results obtained by the authors of the book and are a concise summary of the achievements of other researchers—first of all the results obtained by Aronov et al. [1]. This was necessary, since further references are made to this chapter when substantiating the provisions of methods for assessing the levels of safety and the significance of risks in a new interpretation resulting from the classical reliability theory.

The basic concepts of ensuring safety in the classical reliability theory for technical systems are given. The methods of qualitative analysis and preventive methods of handling failures are considered on the basis of probabilistic and statistical analysis of the safety of complex technical products. The concept of "acceptable risk" is introduced. It is pointed out that it is necessary to ensure the reduction of damage in case of potential accidents and increase the level of safety taking into account the risk of damaging the entities of the operation process. There is a need in an analysis of essential and significant scientific and technological achievements in the field of RT that enabled the creation of highly reliable systems, especially those such as aviation technical systems (ATS), which is necessary in the development of flight safety management systems and aviation activities based on ICAO recommendations and the new provision of Annex 19 [2–5].

1.1 Formation of Methods for Ensuring Reliability and Safety of Equipment as Quality Characteristics

The basis of the theory for the solution of reliability problems (RT) was the probability theory and mathematical statistics.

However, procedures for managing risks of accidents using techniques to ensure the fail-safety of machinery are not well formed.

Kuklev E.A. et al., *Aviation System Risks and Safety*, Springer Aerospace Technology, https://doi.org/10.1007/978-981-13-8122-5_1

In the RT, the concept of "safety" is just a consequence by default, without reference to any international standard. At the same time, a lot of works are already known that cover a new scientific direction, the system safety theory (SST).

In this connection, the main positions of the classical RT and PSA are compared using the "risk" categories.

When performing a large number of reliability tests, a critical analysis of the failure causes showed their significant dependence on the design of products, production technology, and operating conditions (adverse factors in the safety theory). The requirements of American standards have been most fully implemented in the extensive program APOLLO aiming to ensure reliability and safety of spacecraft in the process of their development, production, and ground testing.

Within the reliability theory (RT) [3–10], a safety theory was formed on the basis of the initial assumption that deterministic calculations of the process parameters with the worst-case design basis accident ensure the safety of the facility during operation on the basis of the "if it is reliable, then it is safe" principle [6–8].

Recognition of the probabilistic nature of accidents led to a change in the concepts of safety and to the recognition of such a category of concepts as "*acceptable risk*" and "scenario of events" [9–19] (I. Ryabinin's school).

The concept of the accident occurrence risk in the RT as a universal safety feature determined the development of the probabilistic safety analysis (PSA). A conclusion was drawn that after the failure probability has been calculated, it is necessary to evaluate the consequences of the failure [1, 6, 17].

In this book, certain statements included in the "*system safety theory*" and allowing for adjusting some incorrect results in "rare events" problems [15–23] are taken as indisputable:

- "*safety*" is to be assessed through the "*risks*" category;
- "*accident rate*" *and* "*catastrophes*" depend on the probability of some "*scenarios*" of the development of technical processes.

Three typical variants of the action strategy are known: "*attempt to avoid risk*", which is not always possible, since the impact of hazardous factors is continuous; "*neglection of the risk*", which is not an optimal option, since the damage from accidents can be significant; "*risk management with identification of factors* (*predicted threats*)" under the conditions of informational uncertainty regarding risk situations.

1.2 Basic States of Facilities in the Reliability and Safety Analysis

Such property of the facility as *reliability* is defined in GOST 27.002-89 "Industrial product dependability". Reliability is a property that includes fail-safety, durability, repairability, and maintainability.

For safety analysis, the most important property is *fail-safety*—the property of the object to continuously maintain an operational state for some time or some operating time before a critical failure occurs (according to the RT) (Tables 1.1, 1.2, 1.3 and 1.4).

In the RT documents, safety is considered as a technical and economic property of products; therefore, the basic term is the concept of "damage", which refers to the

Table 1.1 Risks to human health from power plants (deaths/GW(el)/year)

Power plant	Risk nature	Work-related fatalities	Mortality
Coal-fired plant	Direct Remote	0.023–0.44	0–77
Natural coal plant	Direct Remote	0.009–0.018 Low	3.2–22 0.0025–0.017

Table 1.2 Data on emergency situations (ES) that occurred in 1997

Source	Total	ES magnitude			
		Le	Ls	Rs	Fc
Man-made sources	1174	871	298	2	3
Derailments, crashes, and collisions with passenger trains	19	5	14	0	0
Accidents on cargo and passenger ships	34	25	8	0	1
Aircraft accidents	31	20	10	1	0
Major car accidents	151	133	18	0	0
Accidents on main and in-field pipelines	81	39	42	0	0
Accidents at industrial facilities	250	202	48	0	0
Detection of explosives in inhabited localities	66	63	3	0	0
Chemical accidents	96	77	19	0	0
Loss of radioactive sources	28	27	1	0	0
Accidents in buildings of residential and social purposes	304	253	51	0	0
Accidents with life support systems	114	27	84	0	0

ES scale: Le—local element; Ls—Local and sub regional; Rs—region; Fc—federal

Table 1.3 Data on transport accidents in Russia

Accident type	Total, persons	Killed, persons	Injured, persons			
	1994	1995	1994	1995	1994	1995
Roadway accidents	177	184	620	620	1611	1363
Accidents with pipelines	38	48	1	2	2	9
Railway accidents	88	52	47	24	160	60
Aircraft accidents	35	42	402	168	466	319

Table 1.4 Critical vehicle defects

Country	Make of vehicle	Description of a typical critical defect	Vehicle production volume, pcs
Great Britain	BMW AG	In case of overheating, the valve may not work causing damage to the cooling system	205,779
	BMW AG	Insufficiently reliable connection of the brake pedal to the booster	166.46
USA	Chrysler Sebring, Dodge Avenger, Eagle-Talon	Premature wear of ball joints in the front suspension	170,000
	Mitsubishi Galant		294,000
	Nissan Infiniti Q 45 and Nissan Infiniti 130	Defective diode in the generator rectifier assembly	2295 and 14,140
	Chevrolet Silverado, GMC, Chevrolet Suburban, Tahoe, GMC Yukon	Steering defect	About 2,000,000
Australia	Ford Falcon, Fairlane	Loose bolt in the front suspension	53,000

relative characteristic of the state of the object that reflects the deterioration of the quality of the object [1, 6, 8].

Following this notion, two properties of the facility are logically introduced into the RT:

- *facility hazard*, the property of the object characterized by its capability to cause damage;
- *facility safety*, the property of the object characterized by its capability to prevent damage or to limit its magnitude;
- *hazardous state*, the state of the facility characterized by damage exceeding the amount of acceptable damage;
- *safe state*, the state of the facility characterized by damage that does not exceed the amount of acceptable damage.

However, these RT concepts do not hold up against serious criticism of the system safety theory (SST), where "safety" means a "state".

Wherein:

- State of the facility characterized by an unacceptable probability of damage can be called "*emergency state*";
- State of the facility if it can perform its intended functions is called "*operable/operational*".

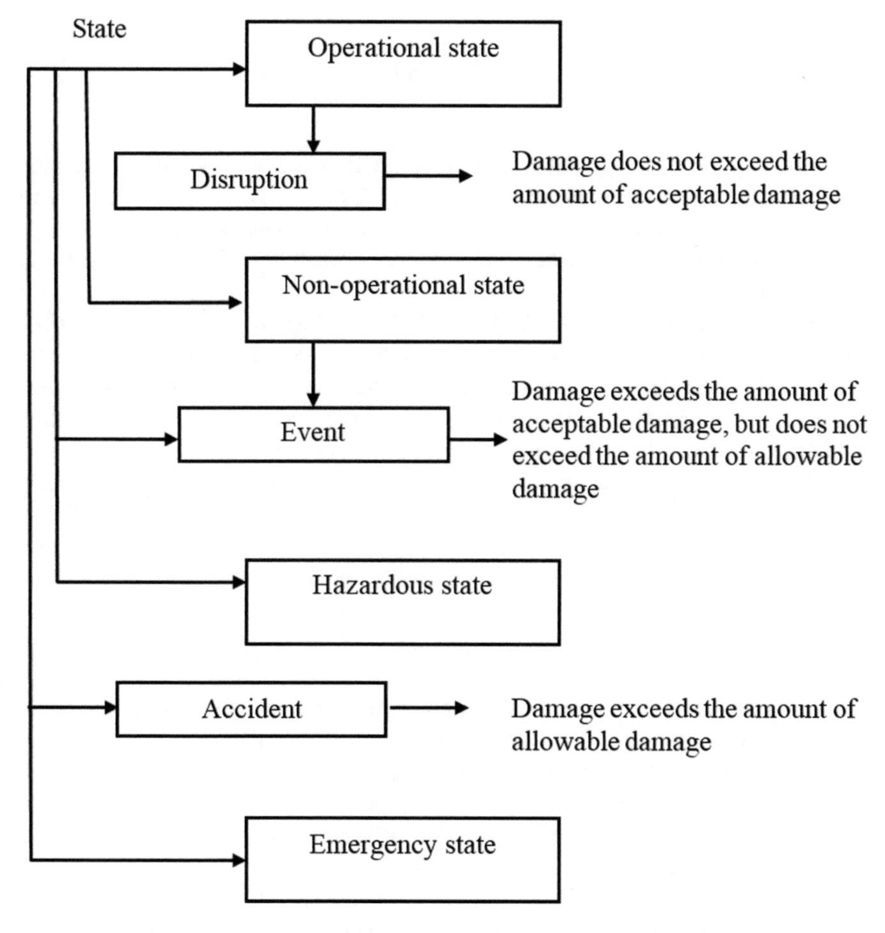

Fig. 1.1 Diagram of main events and states in the facility safety analysis

The classification of events and states of the facility related to safety is shown in Fig. 1.1.

Failures of the facility elements in the safety assessment are characterized by the *criticality of failure* (its severity) as a set of signs of the failure consequences [6, 8, 15]. *But this is the essence of the problem of determining how the magnitude of the risk depends on the factors.*

For non-recoverable systems (up to the first failure), the probability of fail-safe operation $P(t)$ depends on the operational time t, continuous and differentiable.

$$P(t) = 1 - F(t),\ P(t) = 1 - \int_0^t f_0(t)\mathrm{d}t, \tag{1.1}$$

$$F(t) = \int_0^t f_0(t)\mathrm{d}t, \tag{1.2}$$

where $f(t)$ (pdf) of the operating time t before failure is given by the dependences from Table 2.3, where $f_0(t)$ is the sign of the function ($t > 0$). A similar indicator in various forms is associated with indicators such as the probability of flight accident and the probability of a particular situation, the probability of failure to fulfill an intended function.

The safety indicator in the RT is the *criticality C of a failure* as a certain number from a given range (dimensionless scale) from the consequences of the failure.

In this case, the mathematical *expectation of the number of non-desirable events* (failures, NOFs, accidents) that occurred over a fixed time interval T is the average number of flight accidents per 100,000 flight hours (Dispersion, if necessary).

In the classical RT, it is accepted that risk is the most important indicator of safety, since it characterizes the facility in the probabilistic aspect, regarding the facility's "*capability to prevent*" the formation of damage. In the RT, the risk R_p as the value of the "event probability" is calculated by the formula [1]:

$$R_p = P\{\text{damage} > \text{allowable damage}\}, \tag{1.3}$$

where P is the sign of the probability calculation operation. (This is difficult, since one has to consider events with unknown Prdf in the safety theory).

Comment (*by Kuklev*): *In the classical RT, the concept of "risks"* (*by formulas*) *is not in line with the latest international standards* (*e.g., per ICAO*), *but here the concept of risk from Annex 19* (*ICAO*) *is adopted.* Then, one can introduce corrections to the "indicators" of safety.

Indicators of the risk type and criticality in the RT shall be considered as *probabilistic complex universal negative indicators of safety.*

(*Comment of the authors of the book: "direct measurement" is doubtful if there are errors due to small sample volumes*).

1.3 Interrelationship Between the Categories of Reliability, Efficiency, and Safety of Complex Technical Systems in the Classical Reliability Theory

If the product consists of n components, each of which can be in one of two mutually exclusive states—operational and non-operational, then the state of the product is described by the number $N = 2^n$ of incompatible events $H = \{H_m\}$, which constitute the complete group of events in the state matrix H:

$$H = \left\| \begin{matrix} H_0 \\ H_1 \\ \vdots \\ H_n \\ H_{12} \\ \vdots \\ H_{12\ldots n} \end{matrix} \right\| = \left\| \begin{matrix} X_1\, X_2 \ldots X_n \\ \overline{X}_1\, X_2 \ldots X_n \\ \vdots \\ X_1\, X_2 \ldots \overline{X}_n \\ \overline{X}_1\, \overline{X}_2 \ldots X_n \\ \vdots \\ \overline{X}_1\, \overline{X}_2 \ldots \overline{X}_n \end{matrix} \right\| \tag{1.4}$$

where X_m denotes an event of the operational state for the mth component, and $\overline{X}_m$ is a failure event for the mth component i; $\overline{(\cdot)}$ is a sign of logical negation. The state matrix H can be compared the probability matrix for these states $P = \{P_i\}, i = \overline{1,\ m},$ $m = N = 2^n$:

$$P^T = ||P(H_0)P(H_1)\ldots P(H_n)P(H_{12})\ldots P(H_{12\ldots n})||, \tag{1.5}$$

where T is the matrix transposition sign.

Since the states $\{H_m\}$ form a complete group of incompatible states, the sum of the elements of the P matrix is equal to one:

$$P_m = P(H_0) + P(H_1) + \cdots + P(H_n) + P(H_{12}) + \cdots + P(H_{12\ldots n}) = 1,$$

$$\Phi^T = \Phi^T = ||\Phi(H_0)\Phi(H_1)\ldots \Phi(H_n)\Phi(H_{12})\ldots \Phi(H_{12\ldots n})|| \tag{1.6}$$

In the ideal case, using H, P, Φ, one can construct a discrete law of probability distribution for conditional efficiency, for example, mathematical expectation, mode, and median.

(*Comment by Kuklev: Unfortunately, some "mathematical expectations" adopted in the RT are not calculated in the SST because of the rare events problem.*) (Table 1.5).

1.4 Structurally Complex Diagrams of the Technical System and Minimal Cut Sets of Failures

1.4.1 Methods for Assessing Reliability and Quality of Systems

Qualitative analysis of reliability for an individual system includes, first of all, the construction of a structural-functional diagram of the system on the basis of a description of its modules and each mode of its operation. The steps of this analysis are as follows [1, 6, 8]:

Table 1.5 Mathematical expectation and variance of typical distributions

Distribution	Mathematical expectation	Variance
Normal	μ (*1)	σ^2 (**1)
Lognormal	$\exp\{\mu + \sigma^2/2\}$(*2)	$\exp\{\mu + \sigma^2/2\}(\exp \sigma^2 - 1)$ (**2)
Exponential	$1/\lambda$ (*3)	$1/\lambda^2$ (**3)
Gamma distribution	m/λ (*4)	m/λ^2 (**4)
Weibull	$S_0 + (\theta - S_0)\times$ $\times\Gamma\left(\frac{1}{\beta}+1\right)$ (*5)	$(\theta - S_0)^2\left\{\Gamma\left(\frac{1}{\beta}+1\right) - \left[\Gamma\left(\frac{1}{\beta}+1\right)^2\right]\right\}$ (**5)
Maximum values	$S_0 + 0,5776$ (*6)	$1,645\,\theta^2$ (**6)
Minimum values	$S_0 + 0,5776$ (*7)	$1,645\,\theta^2$ (**7)

- analyzing the types and consequences of failures of system elements;
- selection and construction of a structural-logical model of system reliability for each selected criterion of its failure in the form of failure trees;
- *definition of a set of minimal cut sets* to identify weak points in the system.

The *minimal cut set* is the aggregate of the list of events in the system in the form of failures of the facility elements with the following properties:

(a) Their combined occurrence leads to the system failure, i.e., loss of functions.
(b) Any combination of fewer events (including failures) does not lead to a system failure.

The method for determining the analysis of properties (MCF) is given in detail in [24, 25] (I. Ryabinin).

1.4.2 Constructing a "Failure Tree"

When constructing a "*failure tree*", a certain undesirable event (vertex event) is identified, and then, possible causes of its occurrence are analyzed. The "failure tree" makes it possible to analyze the significance of individual elements of the system in a qualitative form with the help of *minimal cut sets of the "failure tree"* (MCF), which is useful for engineering design.

With a known set K of minimal cut sets $e_1, e_2, \ldots, e_k$, the probability of fail-safe operation of the system P can be estimated from:

$$P \geq \prod_{i=1}^{k} p(e_i) = \prod_{i=1}^{k} [1 - q(e_i)] \tag{1.7}$$

where $p\ (e_i)$ and $q\ (e_i)$ are the probabilities of fail-safe operation and failure of the ith cut set, respectively, if the ith cut set includes a single element of the system. If several elements of the system are included in the ith cut set, the probability of fail-safe operation of this cut set is calculated by formula:

$$p(e_i) = 1 - \prod_{i=1}^{n_i} (1 - p_i) = 1 - \prod_{i=1}^{n_i} q_i, \tag{1.8}$$

where p_i and q_i are the probabilities of fail-safe operation (failure) of the ith element included in the ith cut set (i = 1,2, …, n), and $P\ (e_i)$ is the probability of the system failure in the cut set e_i. The condition for applying the statistical method of testing hypotheses is that the measured characteristic X has a distribution of type $N(x) = (a,\ \sigma^2)$; i.e., it is distributed according to the normal law with mathematical expectation a and variance σ^2. Moreover, the hypothesis H_0 is not rejected until the value of X, calculated from n individual values, satisfies the inequality [1]

$$\mathrm{a} - Z_\alpha \frac{\sigma}{\sqrt{n}} < \overline{X} < a + Z_\alpha \frac{\sigma}{\sqrt{n}}, \tag{1.9}$$

where a is the nominal value of the characteristic; Z_a is the value of the critical value of the hypothesis H_0 for a given probability with an rms deviation ("variance") σ of the error; n is the number of individual values of the characteristic, $\mathrm{X_i},\ i = \overline{1,\ n}$.

1.5 Basic Principles of Ensuring Safety of Technical Systems Based on the Classical RT Methods

1.5.1 Use of Safety Barriers to Ensure Safety of Potentially Hazardous Facilities

The *principle of defense*, that is, *defense in depth*, involves the creation of a series of protection levels against possible failures of the facility elements and personnel errors [1, 6, 10]. Accidents occur in the event that all safety barriers fail.

Under certain conditions [1], the probability of fail-safe operation P of an *aggregate* of n barriers is:

$$P = 1 - \prod_{i=1}^{n} q_i. \tag{1.10}$$

A gain in the probability of fail-safe operation is achieved.

1.5.2 Place and Role of Probabilistic Safety Analysis (PSA) in the RT

The PSA methods are intended for quantitative safety assessments, since this method is the basis of safety management, when the decision is made on the basis of reliable facts.

PSA methods are used to determine the risk (at the design and operation stages), to optimize design decisions by comparing several variants of the facility.

According to this method, the criterion of the optimum level of safety is the minimum value of Z, which is the sum of two components: $X(m)$, adjusted safety costs characterized by risk R, and $Y(R)$, direct damage caused by this risk:

$$Z_t = \arg\min Z\{r\} = \arg\min[X\{r\} + Y\{r\}] \quad (1.11)$$

In many cases, the normalization of the acceptable level of risk is based on the ALAP (*as low as possible*) principle in the form of a compromise between the society's requirements for safety and the resources available to provide the required level of safety for standard types of threats.

1.5.3 Identification of Risk Factors

In the RT, the risk assessment is associated with the level of potential hazard for the system as a whole, depending on the severity of the damage and the probability of the hazard, as well as the probability of damage, the frequency, and duration of the hazard for people, the probability of a hazardous situation, etc. These formulations are not strict, there is no concept of a hazardous event, and the methods for assessing the significance of the level of threats and corresponding hazards in the RT have not been developed, although there are PSA methods, in particular those from NASA [12].

Tables 1.6 and 1.7 provide data on the possibility of events from (1.6) (general statistics).

1.5.4 International Standards in the Field of Safety Analysis and Evaluation (PSA) and Comments on Discrepancies in Language

The analysis of international documents [18, 23] identified two important trends. First, *they consider safety as the absence of unacceptable risk* (*ISO/IEC Guide 51: 1990*) or as the capability of a facility (machine) to perform the required functions under conditions intended for its use and specified by the manufacturer, *without any*

Table 1.6 Criteria-based values for risks of fatal outcome as a result of natural phenomena and for some activities

Hazard type	Risk value, 1/h
Natural habitat	10^{-12}
Hurricanes, thunderstorms, radioactive substances in goods	10^{-11}
Typhoons, bites of venomous insects, earthquakes	10^{-10}
Firearms in everyday life, gas explosion in tenement houses	10^{-9}
Diseases, age of 10–14	10^{-8}
Diseases, age of 30–35, cycling, hunting, flying	10^{-7}
Smoking, diseases, age of 60–69	10^{-6}
Aircraft tests	10^{-5}
Horse racing, regular flights on military aircraft	10^{-4}
Participation in racings	10^{-3}

Table 1.7 Criteria-based values for accident risks for some facilities

Facility type	Accident risk, 1/year	Facility type	Accident risk, 1/year
Reactors, core	10^{-7}	Turbogenerators	0
Reactors, main circuits	10^{-6}	Aircraft	0
Rocket and space systems	0	Pipelines	2×10^{-3}

risk of health damage (*standard EN 292-1: 1991, standard ISO 12100-1: 1992*) [2, 3, 7, 13, 14].

This definition does not correspond to the SMM (No. 9859) and the latest recommendations of Annex 19.

The concept of "absence" of unmanageable risk should be removed.

Quantitative risk assessment can be performed by several methods: FMEA (failure mode and effect analysis), FTA (failure tree analysis (events, faults)), HAZOP (hazard and operation analysis), etc.

Standard ISO 14971: 2000 provides a single *risk management* process.

1.5.5 Identification of Main Tasks of Probabilistic Safety Analysis

An important PSA task is to identify the most hazardous scenarios that make the greatest contribution to the risk assessment. The PSA basis is the construction of the "event tree" [1, 24] (Figs. 1.2 and 1.3), i.e., system analysis of the events "branching" from the initial event (IE).

Tables 1.8 and 1.9 provide data on the causes of a number of accidents [1, 8, 10].

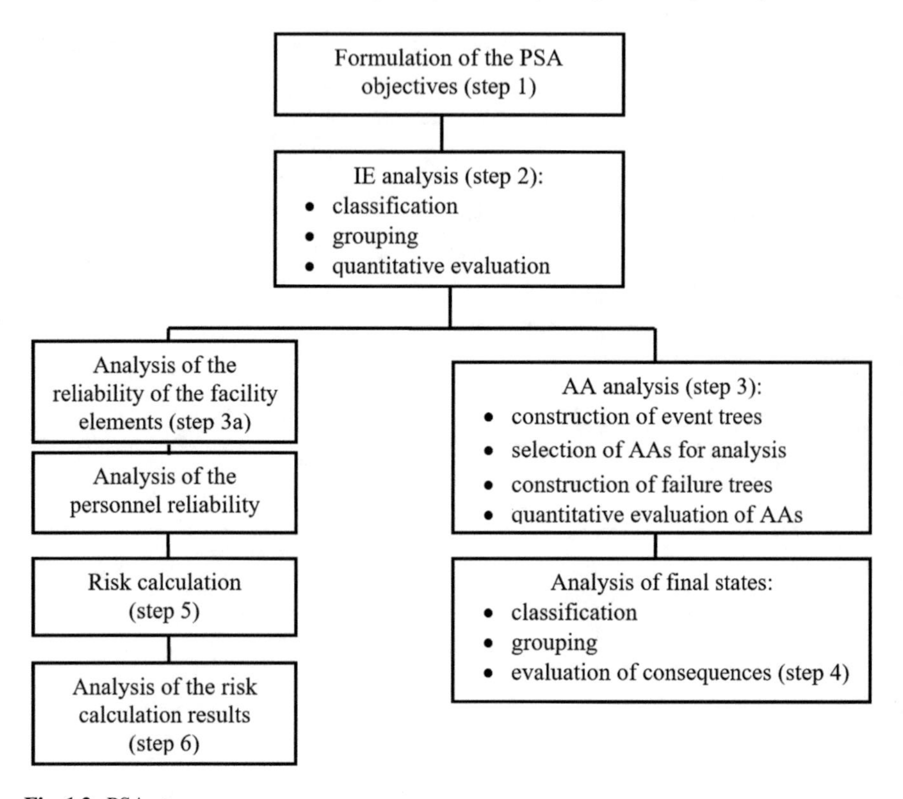

Fig. 1.2 PSA steps

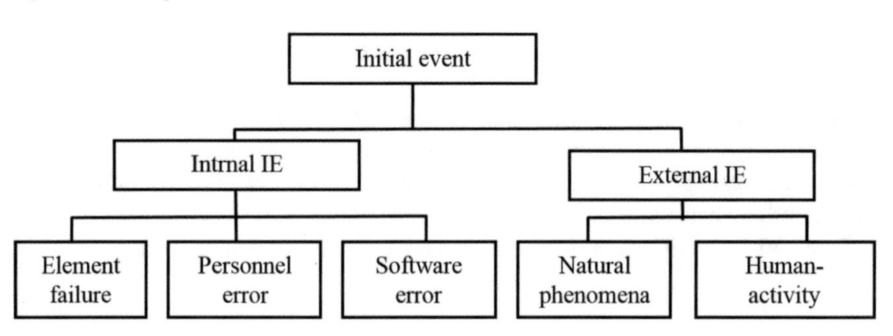

Fig. 1.3 Classification of initial events

Table 1.8 Causes of accidents at nuclear power plants (IAEA data)

Accident cause	Accident fraction, %
Design error	30.7
Equipment wear, corrosion processes	25.5
Operator's errors	17.5
Errors in operation	14.7
Other causes	11.6

Table 1.9 Main causes of accidents involving losses of aircraft (data of the "Boeing" company for all aviation incidents)

Accident cause	Accident number	Accident fraction, %
Flight crew	298	6.1
Aviation equipment	45	9
Maintenance	9	2.0
Weather conditions	20	4.1
Airport services	19	3.9
Other causes	14	2.8
Unknown	83	17.2
Total:	488	100

It should be noted that PSA is a complex and time-consuming task to be solved at the design stage. In [1], it is indicated that the risk calculation is accompanied by a high degree of uncertainty in the final results. It is this undeniable position that is accepted as the starting point for the creation of a new approach to the SST within the RRS doctrine (in Chaps. 2 and 3 of this book).

But, for example, JSC RR [26] suggests ways to "approve or assign" certain probabilities. However, it is known [27] that "probability" as a non-random, clear number, can only be calculated with Prdf and pdf known in advance. Only an acceptable level of probability or risk can be assigned, if there is a reason to do so. But then, the "true" probabilities should be clearly defined, not "*guessed*".

Paradoxes are possibly known, for example, from the experience in creating ATSs (or ATC operation [28]): Acceptable levels of probability for the occurrence of an event are assigned with the order of 10^{-7}–10^{-9}. But from the standards, it is known that at a value of 10^{-6}, this event "almost does not exist" and it is almost impossible to prove the truth of such a value. These circumstances partly explain the relevance of the positions presented in documents [2–5] for civil aviation.

1.6 Analysis of Emergency Sequences When Assessing the Safety Level of Systems Using the PSA Method in the RT

1.6.1 Construction of "Event Trees" in the RT

The safety of facilities is determined by a chain of events, not all of which are failures. Accidents occur as a result of the emergence of special chains from a sequence of elementary events. The "tree" describes many possible ways of the occurrence of an incident or some other event that determines the risks.

For elements with continuous operating time t, the calculated indicator is taken in the form of the probability of fail-safe operation $P\{t)$ for a given operating time [1, 24, 27].

1.6.2 Calculation of Risks in the RT as the Probability of Occurrence of a Negative Event

In the classic RT [1], it is correctly stated that an accident is a class of realizations from n incompatible emergency sequences of certain events.

Then, the total probability [29] of the accident $R(I_0)$ with the initial event I_0 will be:

$$R(I_0) = P(I_0)\sum_{i=1}^{n} Q_i(E_i|I_0) = \sum_{i=1}^{n} P(I_0)Q_i(E_i|I_0), \tag{1.12}$$

where $P(I_0)$ is the probability of the occurrence of the initial event I_0 for some period of time T, for example, one year.

However, it is known from practice that IE is a rather rare event; therefore, it is accepted in the RT (as in all similar works) that the distribution of the IE occurrence probability over time T can be considered a Poisson distribution:

$$P(v = m) = X^m e^x/m, \quad m = 0, 1, 2, \ldots, \; X > 0, \tag{1.13}$$

1.6.3 Analysis of the Results of Risk Calculation in the PSA Method

It is necessary to take into account the possible uncertainty in the evaluation $R(I_0)$ due to the *branching* of the outcomes; therefore, the analysis of the results of the risk calculation becomes more complicated.

Uncertainty of the PSA results is manifested through the degree of variability in the quantitative PSA results in evaluating small values of the probabilities of events that are opposite to the main event. Therefore, confidence intervals for the significance of the estimates should be determined.

The following simple rule is accepted: *Element A is greater than element B if the number of various minimal cut sets including element A is greater than the number of various minimal cut sets including element B.*

It is known [1, 8, 24, 25] that there are two types of sources of uncertainty in the probabilistic safety analysis:

(a) IE frequencies (or the probability of failure of elements) from the probabilistic models used in the PSA cannot be known exactly.
(b) Incompleteness of scenarios in simple hazard models, hence uncertainty in the PSA results.

Unfortunately, this is not taken into account in the RT [1, 6, 10], and incorrect recommendations are given that the following indicators can be a measure of *uncertainty*: *variance, root-mean-square deviation, variation coefficient* obtained in simulation modeling.

Comment (*by Kuklev*): *Simulation modeling is a completely non-working tool in the rare events problem*, as proved by NASA [12, 13].

1.7 Failure Mode Effects and Criticality Analysis (FMECA)

1.7.1 General Provisions of Failure Mode Effects and Criticality Analysis for System Element Failures

Failure mode effects and criticality analysis (*FMECA*) is a group of PSA methods for the preventive analysis of failures and defects in the facility elements that affect the safety of processes (process FMECA).

The FMECA results in corrective measures to reduce the criticality of individual failures and improve the safety of the facility or aviation activities.

Tables 1.10 and 1.11 show the performance characteristics of the technology in question. Also, the estimates of the significance of event frequencies are given that may be important for solving the rare events problem.

The FMEA includes the following quantitative estimates of the failure frequency [1]:

Expected failure frequency	Associated failure probability
Frequent failure	$P > 0.2$
Probable failure	$0.1 < P < 0.2$
Rare failure	$0.01 < P < 0.1$
Very rare failure	$0.001 < P < 0.01$
Improbable failure	$P < 0.001$

The coefficient C_i (criticality) of the ith element of the facility is calculated in the RT as follows:

$$C_i = B_{1i} \cdot B_{2i} \cdot B_{3i}, \tag{1.14}$$

where values B_i ($i = 1, 2, 3$) are taken from the corresponding tables developed beforehand depending on the system type, but, as a rule, $1 < B_i < 10$; $B_{1i} \sim B_u$ is the estimate of the frequency (probability) of the potential failure of the ith element; B_{2i} is the estimate of the probability of detecting the failure (defect) of the ith element before its manifestation on the consumer's side; B_y is the estimate of the ith element failure (defect) consequences severity.

The FMEA diagram (above) and (1.14) already envisage the necessity to transit to the Fuzzy Sets in the SST, since the concept of "rare failure", etc., is fuzzy linguistic variables, and the concepts of "probability" of the failure determine clear measurability of events (the rationale is given in Chap. 3).

Table 1.10 Classification of FMECA types

FMECA types	FMECA objectives	Failure effect
Facility	Analysis of the facility element failure effect	Causes a critical failure of the facility and can lead to an emergency state
Process	Analysis of the operation effect on the final product Analysis of the operation effect on the process	Causes a critical defect in the product. Stops the process and affects the safety of the personnel and the environment

Table 1.11 List of documents establishing FMECA procedures

Document name
FMEA: Fehler – Moglichkeits und Einflus – Analyse, Notwendigkeit, Chance. Voraussetzung. Grundlagen, Volkswagen, 1988
Qualitatskontrolle in der Automobil-industrie, Sicherung der Qualitat von Serieneinsatz. VDA 4., Frankfurt am Main, 1986
MIL–STD–1629A: Procedures for Performing Failure Mode, Effects and Criticality Analysis. Washington, DC 20301, USA, 1980
IEC Publication 812. Analysis Techniques for System Reliability—Procedure foT Failure Mode and Effects Analysis. IEC, Geneve, 1985

1.7.2 Effect of the Failure Criticality on the Safety State of the System Processes

In the RT (according to [1, 6, 8]), the criticality of events is evaluated in a very narrow sense, which does not reflect modern approaches, for example, in the SST. This analysis is made to show the need to develop a theory of the System Safety (SST) on the basis of the ICAO methodology for calculating risks [2, 5, 30].

The idea of analyzing the criticality of an operation is to take into account the following factors: the frequency of the defect due to the loss of the operation accuracy and the probability of identifying this event and the consequences of the failure.

In the general risk theory (in the cited paper [31] ([1])—see below) the use of the term "risk" is analyzed for various areas in risk assessment and management.

Description of risk phenomena means, for example, the following (for a driver): undesirable probabilities (hazards, risks) (1) to hit a traffic jam; (2) to have a car accident; (3) to be attacked by crime figures, etc. The most common approach is based on the probability theory, but fuzzy and interval mathematics is also considered. It is this probability that is called a risk by some authors, but they also take accidental damage (hazard severity) into account. The risk assessment is reduced to a statistical evaluation of the parameters, characteristics, and dependencies included in the model. Here, the following is assumed: Risk is a measure of a quantitative multicomponent measurement of a hazard with the inclusion of the amount of damage from the safety threats, the probability of these threats, and the uncertainty in the magnitude of the damage and probability. It is assumed that the distribution function belongs to one of the known families of distributions, that is, normal (i.e., Gaussian), exponential, or some other.

However, this assumption is usually not well founded, since it is difficult to find a priori *pdf*, that is why other ways are needed to solve the problems under consideration (Table 1.12).

1.7.3 Examples of Known Catastrophes

Natural Catastrophes in the Krasnodar Region that have occurred for several years in a row due to floods and breaches of protective dams in water storage reservoirs are the most typical example of inappropriate design decisions made only on the basis of RT methods without taking into account the SST recommendations. (The usual wording is the following: "*Such events are rare; therefore, it is inadvisable to create protective structures*").

"Fukushima" It can also be assumed that the catastrophe with a nuclear power plant in Japan followed a similar scenario (mentioned as "Fukushima" in the media). If one does not take into account the special circumstances associated with business

[1]UDC 519.2:330.:658.5. LBC 65.050 65.290-2. APPROACHES TO THE GENERAL RISK THEORY. A. Orlov, O. Pugach (Bauman Moscow State Technical University).

Table 1.12 Assessment of the failure significance

Expected failure frequency		Failure category		
	I	II	III	IV
Frequent	*A*	*A*	C	C
Probable	*A*	*A*	*B*	C
Rare	*A*	*B*	*B*	*D*
Very rare	*A*	*B*	*B*	*D*
Improbable	*B*	*C*	*C*	*D*

Note by V.G. This is true, but here only the aspect of industrial safety by fact F1 ("design") is considered, which is analyzed below in paragraph 1.8—risk models by Bauman MSTU

in nuclear energy, this catastrophe can also be classified as a rare event, the analysis of which did not include some dangerous constructive and natural factors that were very rare in terms of the "possibility" estimate (and even more rare in terms of the "probability" estimate).

(A quotation from the source [11] is given here).

"Fukushima" Taught Lessons: The consequences of this disaster in Japan were discussed at the International Economic Forum.

- The tragedy in Japan can change the whole picture of demand", said Nobuo Tanaka, head of the International Energy Agency. Countries are thinking about alternative sources of energy. However, experts agreed that there is no need to abandon nuclear energy, but it is necessary to pay attention to safety issues.
- By 2025, the need for energy will grow by 65%; alternative sources will not be able to withstand the increased workload", said Bill Richardson, New Mexico Governor.

In this regard, the Rosatom CEO Sergey Kiriyenko said that it is necessary to change international legislation and introduce general global standards. "In addition to the technology of nuclear power plants, the experience of the operating organization and its qualification are also important". Moreover, according to Kiriyenko, we need safe development of nuclear energy, and this is ensured if we switch to new technologies that already exist. More than 80% of the experts voted for the necessity of complying with the recommendations of the International Atomic Energy Agency (IAEA)" (translated from Russian).

1.8 Conclusions

1. The classical RT provides the fundamental pillars of the system safety theory with the use of a calculating apparatus and tools for evaluating safety indicators, such as the PSA. This method is used in industry, transport, and nuclear energy

domains. However, the results of estimating the probabilities of "*rare events*" are proposed to be found in the "*fuzzy region*" of confidence intervals. This is incommensurable with the real-life systems due to the lack of reliable statistics on emergency and catastrophic situations.
2. The concept of risk in the RT is formulated as the "probability of the occurrence of events" with negative outcomes (damage). A similar wording is adopted in insurance and in the financial sphere in order to assess the "*average risk*". The proposed wording does not provide correct solutions to situations with events of "near-zero" probability of occurrence.

New approaches are needed that take into account the ICAO recommendations (per Annex 19).

References

1. Aronov IZ et al (2009) Reliability and safety of technical systems. Moscow (in Russian)
2. Annex 19: C-WP/13935—ANC Report (March 2013), based on AN-WP/8680 (Find) Review of the Air Navigation Commission, Montreal, Canada
3. SMS & B-RSA (2008): "Boeing", 2012
4. Amer MY (2012) 10 Things you should know about safety management systems (SMS). SM ICG, Washington
5. SMM (Safety Management Manual) (2012) Doc 9859_AN474—Doc FAA
6. Aronov IZ (1998) Modern problems of safety of technical systems and risk analysis. Standards and quality, no 3 (in Russian)
7. British Standard (1992) Quality management and quality-assurance. Vocabulary. VS EN ISO-8402
8. Volodin VV (ed) (1993) Reliability in technology. Scientific-technical, economic and legal aspects of reliability. Blagonravov Mechanical Engineering Research Institute, ISTC "Reliability of Machines"—RAS, Moscow, pp 119–123 (in Russian)
9. Risk management—Vocabulary—Guidelines for use in standarts. PD ISO/IEC, Guide 73: 2002. B51: 2009
10. Kochetkov KE, Kotlyarevsky VA, Zabegayev AV (eds) (1995) Accidents and catastrophes. Prevention and mitigation of consequences. Study Guide. In 3 books. Publishing House ASV, Moscow (in Russian)
11. Waitmann M (2011) UN: Japanese nuclear power plants are not ready for a tsunami. www.metronews.ru, 02.06.11 (in Russian)
12. Probabilistic Risk Assessment Procedures for NASA Managers and Practitioners—Office of Safety and Mission Assurance NASA. Washington, DC 20546—August 2002 (Version 2/2)
13. McCarthy J (1999) U.S. Naval Research Laboratory; Schwartz N., AT & T. Modeling Risk with the Flight Operations Risk Assessment System (FORAS). ICAO Conference in Rio de Janeiro, Brasil
14. Documents of the 37th ICAO Assembly (Oct 2011, Montreal)
15. Severtsev NA (ed) (2008) Kuklev EA "Fundamentals of the system safety theory", RAS Dorodnitsin CC. Moscow, pp 175–180 (in Russian)
16. Gipich G, Evdokimov V, Kuklev E, Mirzayanov F (2013) Tools: identification of hazard & assessment of risk. Report at the ICAO meeting "Face to Face". "Boeing" Corp., Washington
17. Severtsev NA (ed) (2005) Issues of the system safety and stability theory, issue 7. RAS Dorodnitsin CC, Moscow (in Russian)

18. Accident Prevention Manual (1984) Doc. 9422-AN/923. International Civil Aviation Organization
19. Risk management—principles and guidelines. Standards (Australia)—AS/NSZ ISO 31000: 2009
20. Smurov MY, Kuklev EA, Evdokimov VG, Gipich GN (2012) Safety of civil aircraft flights taking into account the risks of negative events. J Transp Russ Fed, 1(38):54–58 (in Russian)
21. Malinetskiy GG, Kulba VV, Kosyachenko SA, Shnirman MG et al (2000) Risk management. Risk. Sustainable development. Synergetics. Series "Cybernetics" Nauka, 431p. RAS, Moscow (in Russian)
22. Documents of the ICAO High-level Conference (SMM). March 2010, Montreal
23. Risk management—vocabulary. ISO Guide 73: 2009 (E/F). BSI: 2009
24. Ryabinin IA (1997) Reliability, survivability and safety of ship electric power systems. Kuznetsov Naval Academy, St. Petersburg (in Russian)
25. Belov PG (1996) Theoretical foundations of system engineering safety. GNTP Safety MIB STS, Moscow (in Russian)
26. Procedure for determining the acceptable level of risk (URRAN). JSC Russian Railways standard STO RR 1.02.035-2010. Moscow, 2010 (in Russian)
27. Prokhorov PZ, Rozanov YA (1987) Probability theory (basic concepts, limit theorems, random processes)—"Science"—FM, Moscow (in Russian)
28. Fujita M. Frequency of Rare Event Occurrences (ICAO collision risk model for Separation minima). RVSM. ICAO, Doc. 2458. Tokio: EIWAC 2009
29. Sakach RV, Zubkov BV, Davidenko MF et al (1989) Sakach RV (ed) Flight safety. Textbook. Transport. Moscow, 239 p (in Russian)
30. SMM: Doc. 9859-A/N460. ICAO, 2009
31. Orlov AI, Pugach OV (2011) Approaches to the general theory of risk. RFBR Grant-2010. Bauman MSTU. Moscow (in Russian)

Chapter 2
New Doctrine "Reliability, Risk, Safety" for System Safety (Flight Safety) Assessment on The Basis of the Fuzzy Sets Approach

2.1 New Doctrine for Assessing Safety of Structurally Complex Aviation Technical Systems Using Fuzzy Subsets

The new doctrine "Reliability, Risks, Safety" (RRS) is a system of views and provisions that makes it possible to create a "bridge" for linking the classical reliability theory and the system safety theory in order to solve system safety issues based on the methodology of calculating risks in the "rare events" problem (according to ICAO) using the Fuzzy Sets approach.

The main traditional apparatus for analyzing and assessing system safety within the RT is the PSA (probabilistic safety analysis or probabilistic risk assessment tools, as described in the primary source, according to NASA and IAEA) method [1–3]. In the RRS, risk identification tools are described in the system safety theory (SST).

The term SST was first introduced in the Russian Academy of Sciences (Dorodnitsyn Computer Center) by Severtsev [4]. However, the methods for solving "rare events" problem remained classical, as in the RT (using the PSA). In the RRS, as mentioned above, a *different approach was adopted*, although the content of theoretical and applied issues is quite close [5–8].

In the RRS, the main link, which allows to correctly combine "reliability" and "safety" into a single whole, is "risks" in the new RAS interpretation ("risk is a measure of hazard") [9].

In the RRS, unlike the RT (and SST, unlike the PSA, respectively), it is assumed that "rare events" such as "accidents" and "catastrophes" have the "near-zero" probability of occurrence of events, whereas in the PSA [10], *nevertheless*, methods are proposed for calculating very small probability values: of the order of 10^{-6} and less; it is difficult to make these calculations reliable. Therefore, such calculations can be ineffective in solving the rare events problem [11–13].

In this regard, methods for assessing safety indicators are different: In the RT (PSA), these are Prdf, pdf, and confidence intervals [10, 14–16]; in the RRS (SST), these are Fuzzy Sets methods [13, 16–18].

Kuklev E.A. et al., *Aviation System Risks and Safety*, Springer Aerospace Technology, https://doi.org/10.1007/978-981-13-8122-5_2

Below, the general and operational provisions of the RRS and SST are presented separately: the concept, the rationale for the approach, the general conclusions and the results of comparing alternative approaches regarding the definition of safety indicators, characteristics of the procedures and algorithms that determine the objective of the tools for risk management in the SST (RRS) [8, 16].

2.2 Multicriteria Estimation of the Complex Quality Index on the Tuple of Parameters

2.2.1 Multicriteria Index and Alternative Methods

The tuple of parameters includes “reliability and safety” indicators. The task is that, according to the ICAO (and NASA) concept to ensure the technical and economic efficiency of the systems, it is necessary to use indicators like “the system is hazardous or safe, i.e., not hazardous”, but taking into account the requirements for “convenience” of operation. This formulation of the problem is reduced to the problem of obtaining *multicriteria* estimates. Such multicriteria tasks can be solved traditionally: numerically with the use of high-speed computers. But in the problem of ensuring safety, one has to consider a complex of the tuple type from sets of mathematical objects that do not form a topological space. In such cases, there are difficulties in applying “linear programming”, “gradient computational methods”, variational calculation, etc. However, various indicators need to be used to at least provide quite formal and irreplaceable requirements stemming from the general modern doctrine of technology development in various spheres of activity, for example, that of the world aviation community.

Promising recommendations were received at the Berkeley scientific school and also announced in civil aviation of Russia. The paper presents examples of solving similar problems based on “fuzzy programming” within the framework of the ***Fuzzy Sets*** approach.

Complex indicators suitable for use in SMSs on the main tuples of sets of indicators have the following form:

$$K_{\sum} = \left\langle K_1, K_2 | \sum\nolimits_0 \right\rangle. \tag{2.1}$$

The following generalized indicators were introduced in (2.1): efficiency, reliability, safety, vulnerability, acceptability, economic feasibility, ergonomics, competitiveness, etc. There are complex dependencies of safety indicators: K_2 for safety management processes and K_1 for aircraft construction reliability parameters. The set of numerical values of the parameters in (2.1) is quite large. Here, the intervals of variation in the indicators of the first type are close to one (for the probabilities of events under reliable operation) or too large numerically (e.g., the cost of an aircraft is several million dollars). The probability of rare events (in highly reliable systems) is "nearly zero". It is practically impossible to create a topological space for tuples of such sets. Almost all variables in (2.1) are fuzzy and are specified on the basis of the classifier of uncertainty types from Table 3.1 (above).

From this, it follows that the traditional probabilistic concept of "risk" is unacceptable in view of the rare events problem: There are no known pdfs and CDFS, and no analytical forms or histograms, and the "statistics" on rare events is unreliable.

It is necessary to create an SMS based on other provisions (without PSA) under challenging conditions: The existence of mathematical expectation (m.e.) and the variance (D_q) of the quantities is not required because there is no information on the average time of the process bifurcation.

That is why we have to switch to alternative methods for finding estimates for (2.1) ***based on the Fuzzy Sets approach.***

Here, a general classifier of uncertainty types can be used (G. Malinetskiy and Table 3.1). When searching for chains leading to a catastrophe, scenario analysis (RAS ICS) and the search for "vulnerability points" and "vulnerability windows" are applied using the new "risk formula" presented in the form of SST provisions studied in this book. It is under this scheme that the ICAO (NASA) risk analysis matrix "works".

2.2.2 Main RRS General Provisions

Based on the results of the factor analysis of situations such as catastrophes regarding engineering and transport, as well as the environment and the financial sphere, the following features were revealed:

(a) The RT methods (under the classical approach) are unsuitable for analysis of catastrophic situations with rare events due to the lack of stable (reliable) statistics on the events of the type considered.
(b) Probabilistic indicators in the RT taken to assess the level of system safety prove to be unreliable due to the lack of clear (exact) laws for Prdf and pdf.
(c) It is inappropriate to introduce characteristics such as the average time to the catastrophe, to the accident, as is customary in the RT, when determining statutory indicators for the average operating time (the average system operability time, for example, to the first failure), due to the insufficient volume of events with severe consequences; such an indicator has absolutely no practical importance in "rare events" problems.
(d) The rare events problem is considered important enough in the system safety theory and is recognized by ICAO (and also by the RAS: N. Makhutov, G. Malinetskiy, I. Aronov, V. Kashtanov, N. Severtsev et al.).

In this regard, the tools for assessing the system safety based on the ICAO "risk management" concept use two methodological approaches:

(a) When assessing the safety levels of systems with "rare events", it is advisable, as far as possible, to abandon "pure RT methods" [10, 15, 19] and proceed to the methodology of calculating risks, for example, per ICAO. Moreover, within the framework of the proposed approach, it is possible to properly take into account both the measure of randomness for "rare hazardous events" and the corresponding consequences of critical events in the form of harm and damage [20];

(b) The traditional approach to assessing the risks of consequences within the concepts of "*risk—probability of negative consequences*" recommended in the RT (and in the PSA) is ineffective ("probability of events with negative consequences") and inaccurate for events occurring with a "near-zero probability". Attempts to overcome the difficulties of the PSA used in the system safety (SS) theory do not yield effective results [1] due to the existence of the "*heavy tails*" problem for pdf, since it is inappropriate to discard these tails when assessing possible catastrophic situations with significant damage and severe consequences for society [9, 12].

2.2.3 General Methodical RRS Recommendations on the Development of Tools for Assessing Risks in Systems as "Measure of Hazard"

When assessing safety of technical systems with complex structural connections of reliability elements (elements that provide functional properties), alternative methods of calculating risks should be used, and the use of probabilistic indicators of the properties of "rare events" should be avoided. Here, the PSA method is considered as a basic one, but alternative for solving the rare events problem under correctly specified conditions.

The sequence of operational procedures in the SST (according to the RRS) is as follows:

(a) Apply the RRS and SST doctrine using approaches based on Fuzzy Sets with fuzzy (with *fuzzily measurable*) indicators of randomness and uncertainty of consequences of rare events.
(b) It is necessary to *avoid* the concept (term) "*risk is a probability* … etc." in the general case of assessing the system safety and adopt the RAS definition "*risk is a measure of hazard*" and, more precisely, the RRS definition "risk is an amount of hazard with a fuzzy measure depending on the fuzzy measure of the predicted possibility of a risk event in systems with a "near-zero probability".

The alternative physical interpretation of the *term "risk"* is as follows: "*risk is a hazard with a fuzzy measure of the amount of hazard* expressed in fuzzy measures such as "more", "less", "rarely", "frequently", and "frequently" without the word "probability", but with using the concept of "likelihood" (realistic possibility, etc.).

(c) To assume that rare events (in the RRS and RT) are immeasurable events with fuzzy measures of randomness and consequences.

NEW DOCTRINE RRS

Rare events: *accidents, catastrophes*

AVIATION TRANSPORT
AN-12 aircraft on rice champ

SEA TRANSPORT
«Costa Concordia» on reefs

Different transport: the same results-catastrophes
(Rare evens-with probability almost "Zero",
Reasons: absence of proactive risk control)

Fig. 2.1 Examples of rare event: catastrophes in civil aviation and on the sea

On the basis of the listed positions, to accept as an initial approximation the agreement on the recognition of the RRS, which allows for correct elimination (and removal) of the contradictions between various indicators of the system quality—"*reliability*" (consumer property) and "*safety*" (an indicator of the state of systems exposed to various adverse factors).

At present, only two types of catastrophes can be distinguished quite clearly in science and technology domains:

type 1 "Catastrophes in synergetic systems" manifested in the form of bifurcations of processes in homeostatic structures (according to V. Arnold and G. Malinetskiy) [9, 21];

type 2 "Catastrophes in technical and polyergatic systems" with the loss of functional properties due to "failures" caused by various intrasystemic and external factors. So far, catastrophes of type 1 have been studied only regarding systems with small dimensions.

The *PSA and RRS cover catastrophes of type 2* that are generated in multidimensional systems defined mainly on the Boolean lattice of the system structure and studied in scientific directions in I. Ryabinin (St. Petersburg) [22], Severtsev [4], and Malinetskiy schools ("*heavy tails*", pdf) [9].

An example of the "rare events" problem is shown in Fig. 2.1 (the An-12 crash—from G. Gipich's archive; the accident with the "Concordia" liner on the reefs—from Internet sources).

The disaster with flooding of the terrain and residential quarters due to unexpected "rain showers" is similar and just as difficult to explain from the RT position. These events can be classified as "rare". In speeches of RT specialists on television, it was stated that a situation like this was envisaged at the design stage, but in view of the rarity of the phenomenon, the risk (according to the probability theory) was insignificant, so the costs of compensation for possible damage were not provided for.

2.2.4 The Main Problems of the Classical RT

*R*are events problem (in civil aviation, according to ICAO); the *uncertainty of the values of Prdf, pdf in the "tails" of the distribution of rare event probability*; lack of stable statistics on rare events (Fig. 2.1).

The existence of these problems was noted by many authors: *M. Fujita, R. Islamov, I. Aronov, A. Orlov [*10–12, 23*].*

Taking into account the analysis of such situations, the methodology, procedures, algorithms, and tools for assessing the risk level and risk management in complex systems for maintaining and ensuring safety are given below.

According to ICAO, "Safety (ISO-8402) is "the state in which the possibility of harm to persons or of property damage is … below… an acceptable level" [14, 24].

"Safe state (SS) is the state of facilities and complex systems, in which adverse events with negative consequences (with damage) can occur, but the level of damage (or the value of risk) does not exceed an acceptable value".

The high reliability of technical systems by probability does not mean that system is safe (it is not related to safety), since in any highly reliable system, a "*residual risk*" is present; i.e., there is a possibility of occurrence, albeit very rare, of an event with very great damage if no measures of proactive control have been taken to manage the possible state of the system.

The "*residual nonzero risk*" of the ATS production is due to design features and technologies and is a sign that "*a catastrophe may occur*".

2.2.5 Possible Ways of Assessing System Safety Indicators with Risk-Based Methods

The provisions of the new RRS doctrine adopted in the SST made it possible to create a methodological base for the correct transition from the PSA provisions to the methodology for assessing system safety and risk levels on the basis of "hazard models" in the form of chains for events from the σ-algebra of the probabilistic space (see below in Sect. 2.4).

At the same time, the SST introduces fuzzy indicators of *integral risk as a measure of the system hazard amount.*

Formally, the RT can simply assign requirements on the basis of "common sense" to safe operation of a highly reliable car in adverse conditions (fog, ice, speeding, complex road relief, etc.). Practical recommendations (seat belts, braking system, choice of headlights, speed limits, road monitoring, etc.) are defined in the form of requirements—"*passive safety*", according to I. Aronov [10, 20].

Solutions of *safety assessment* issues ***start*** in the RT ***at the stage of assessing the "consequences from failures"*** (from functional failures (FFs) [25, 26]) and of assessing risks of losses based on the PSA methods [20, 27, 28], which leads to complex solutions in the *rare events problem.*

Comments on the RT PSA provisions are as follows. The values of the FF probabilities lie in the region of small numbers (the FF probability ~10^{-5}–10^{-12}) and cannot be determined reliably [12] (Kh. Kumamoto). In the SST, the key point is to determine physical causes of the FF occurrence and to assess consequences from accidents with proactive risk management regarding possible catastrophes taking into account risk factors that are significant in terms of damage.

Three basic SST postulates supplementing the classical reliability theory (*RT*):

1. "*The catastrophe is designed in the system and is only waiting for its manifestation*" (J. Reason. from "Massachusetts", author of the theory of chains) [27].
2. The "*residual risk*" reflects hidden (and explicit) errors in the design and production of equipment, as well as possible hidden conditions for the existence of a hazard due to the manifestation of external disturbing and systemic operating factors during operation.

The list of such factors includes threats of the following type:

- *Threats of type 1a*—incidents (sources of danger) actively detected and included in the airline DB on the basis of statistical data;
- *Threats of type 1b*—proactive sources of danger in the form of assumed possible incidents, pilot errors (flight technical error), manifestation of environmental characteristics depending on the factors identified;
- *Hazards*—possible proactive (*predicted states*), hazardous events by threats (risks) introduced, for example, by ICAO, in particular in the form of alternative No. 2 from Table 08\01–ICAO—2007 [27].

The new doctrine formalizes RT and SST theoretical provisions and working tools through the risk category when switching from standard reliability indicators [10, 14, 15] to safety levels expressed through indicator forms of the risk values or by means of some physical indicators proposed by ICAO and NASA [1, 20, 24, 27, 28].

Catastrophes are defined as complex composite events as scenarios of failure sequences in the form of paths leading to a catastrophe along J. Reason's chains.

2.2.6 Relation of Some Parameters from RT and SF into SST

The structure of the chains reflects the properties of the set of "minimal cut sets of failures" [10, 22], but more fully than in the RT.

The functional module of the relationship between the provisions of the classical RT and the SST is shown in Fig. 2.2.

The structure of this module (Fig. 2.2) follows from the relationship between the RT and the SST provisions discussed in Table 2.1. The key point here is to establish the limits of the applicability of the classical RT provisions regarding the study of safety system facilities in fuzzy subsets.

New Doctrine of Civil Aviation Mobility

GENERAL DOCTRINE
"Probability, Risks, Safety"
New doctrine module (for Civil Aviation)

A. Parameters of QMS & SMS
Based on Real Technology Product in CA:
- Technical Object (construction – F1);
- Danger´s Factors of Hazards (F2);
- Data Bases (F1, F2);
B. Distributions for relation according to ICAO:
- **QMS** (High Priority – RT: risk named, as probability):
- **SMS** (Second level – SST: risk named, as measure of danger of Rare Events)

Fig. 2.2 QMS&SMS composition

Table 2.1 Corrected ICAO table (08/01) monitoring the decrease in the risk significance (severity) (measures of hazard)

Risk area	Common threat (hazard sources)	Hazards: description	Consequences (damages from risk events)	Assessment of the event occurrence risk	Existing measures to reduce the risk and the risk index	Measures proposed for further reduction of the risk and the risk index
1	2	3	1–3	2–3	4	5
Example OPS/01	All-weather flight operations at the aerodrome, where one of the two parallel runways is closed due to repair works (an example not associated with this exercise)	Aircraft takeoff and landing on a closed runway (an example not associated with this exercise)	• Midair collisions; • Runway excursion; • Collisions of aircraft on the runway; • Collision of an aircraft with a vehicle on the runway	– High risk (more than the acceptable risk); – Significant risk (unacceptable in terms of the severity of the consequences); – Risk as a rare event (the source is after the first risk event)	1. A NOTAM was issued to inform users about the repair works 2. ATIS (aerodrome repair service) 3. The airport layout is published in the AIP 4. New signs and lighting system	1. Ensure that airline controllers and flight operations managers inform flight crews about the risk of occupying the closed runway by mistake 2. Make sure that the flight crew is notified of the airport layout effective at the moment 3. Mandatory issuance of a NOTAM concerning the closed runway

Notations:

RT	reliability theory;
FS	flight safety;
FR	functional reliability;
FF	functional failures;
*F*1, *F*2	factors (characteristics) in groups:
1	design safety;
2	operational safety;
IS	industrial safety of three types in the MIC:
ILS	integrated logistic support;
QMS	quality management system;
MSG, MEL	replacement strategy for maintenance and repair on the basis of the health monitoring method;
R	risk event: risk assessment (multiple value elements, integral value of the risk (risk), scalar as a function of the set.

Geometrical illustrations of some SST provisions are given that reflect "common sense" in the technical sphere in assessing the consequences from functional failures for aircraft. The basic idea is that even with high reliability of systems and products (*and with a low residual risk*), it is necessary to develop methods for assessing operational safety by determining fuzzy risk values—"*not by probabilities*".

Example 1 The procedure is considered that is adopted in the railway transport to check rail car wheels. A case is analyzed when the aircraft is operated as an air taxi (the aircraft has only one engine) in connection with the assessment of risks with high structural reliability of transport (Table 2.2; Figs. 2.3 and 2.4).

Examples 2 Assessing the significance of risk levels by hazard factors in highly reliable systems

Table 2.2 Dangerous physical situations with "Gzhel" aircraft

Events (hazard chains):	**Control**:	**Risk assessment**:
(a) Engine failure $\varphi_1 = 1$ (b) Gliding descent toward the city $\varphi_2 = 1$ (c) Fall on the quarter The damage is great, the probability of falling is "near-zero", but the "risk" is high ***Decision by the "risk", not by the RT*** "Gzhel" to be replaced by a two-engine aircraft	1. RT: the catastrophe remains, but is delayed to "infinity" 2. SST: (risk) The risk is high, remove factor $\varphi 2$ ***"Gzhel" taxi not to be used***	(a) Not hazardous, since when the engine fails, there is no large-scale "catastrophe" (b) The risk is "low"—"Gzhel" is suitable (c) ***Aircraft engine works on propane gas***: – Risk assessment: the catastrophe is designed by the factor "gas freezing" ***Solution: to prohibit (not permit) the introduction of gas engines in civil aviation***

A. Tapping the wheels of railway cars at the station
"SEARCH OF CRACKS", failure of 10^{-10}
– high reliability

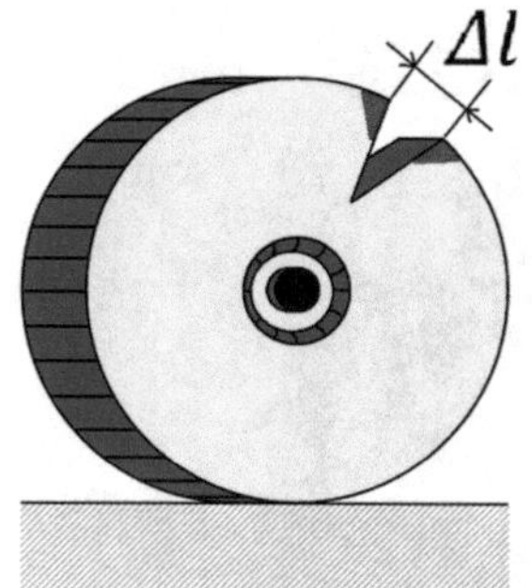

a) $\Delta l < \Delta l^*$ – low risk, traffic permitted
б) $\Delta l \geq \Delta l^*$ – "accident" risk is high
в) Management and impact:
- replace the car
- prohibit the traffic

B. "Roof caving" ("Aquapark", "Subway dome" on 2 supports, brackets) (Catastrophes)

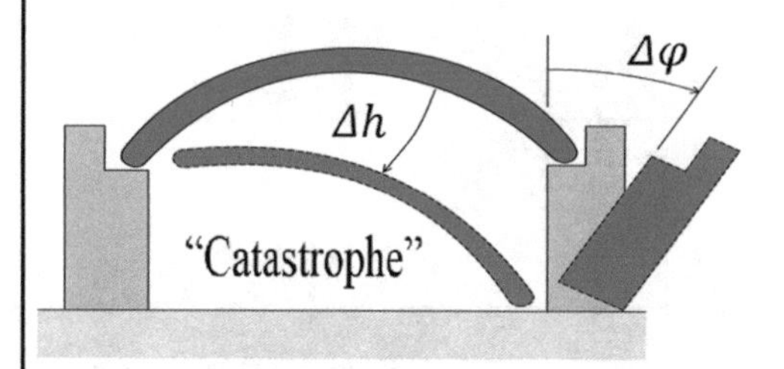

a) Structural risk:
- Bad support (TSNIP);
- Insufficient number of supports.

б) Management (predictive):
- Additional support;
- Welding – "frame".

Fig. 2.3 Critical situations caused by risk factors

Postulate 1—in the SST: The "catastrophe" is designed in the system and is "waiting for its manifestation" at a random time.

Postulate 2—in the RT: "If it is reliable (failure of $\approx 10^{-6}$), then it is SAFE", which is incorrect, because the safety level is not defined by any standard, for example, ISO-8402.

Algorithm for assessing the risks of incident occurrence for an air taxi on the basis of a single-engine aircraft (Gzhel type) as a list of procedures.

2.3 Generalized RT and SST Provisions in the RRS

The risk definition doctrine and the methodology for assessing safety and maintaining the safety level proposed here are based on solving the rare events problem following the physical interpretation of the methodology for calculating risks in highly reliable systems. Above, principles were described in regard to building hazard models in systems based on the study of their state change processes under the influence of

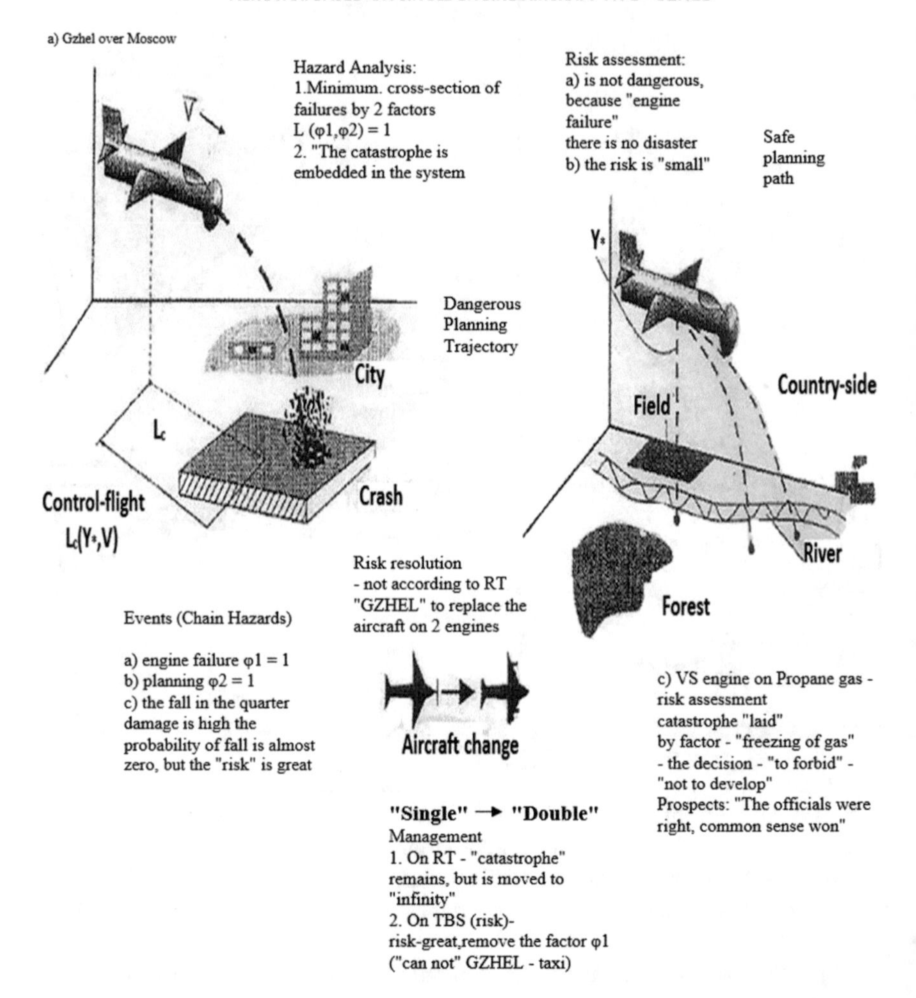

Fig. 2.4 Examples of different risk levels proposed for taxi "Gzhel"

external and internal (systemic) adverse factors. This is equivalent to the procedures for analyzing "event trees" in the RT according to the FMEA standard [29, 30].

From these positions, the mathematical theory of risks and safety is extremely simple and reflects the "common sense" of natural physical representations, concepts, and definitions that are relevant to the concept of "hazard" [31].

Conversely, in the PSA, the mathematical problem and the corresponding theory are somewhat unclear, and practical problems are difficult to solve, as the probabilistic nature of the studied phenomena in assessing the consequences of failures as events opposite to the main event is unclear and distorted in comparison with the original concepts and representations due to insufficient statistics [32–37].

2.3.1 *Interpretations of the Initial Concepts of Risk on the Basis of the Games Theory (Differences in the Classical RT and SST Concepts)*

From the games theory (France), it follows that **risk** is an assumed measure of the possibility of loss—not that loss that has already arisen, but the loss that can be expected. Loss (or win) can be assessed in advance if one decides to participate of a possible loss (in rubles, dollars, etc.). Thus, the risk characterizes some *hazard that there will be some losses* in a hazardous situation with the predicted (expected) risk. A hazardous situation can be defined as the state at the time the game begins, when the decision to "play" is accepted and the events are recognized as possible (Table 2.1).

Chance is luck, happiness, **win**.

Further, from the interpretations of various situations in everyday life and the games theory, it is known (generally accepted) that *risk is most often assessed* taking into account the magnitude of possible or supposed losses—*fuzzily, i.e., in fuzzy terms* [32, 34, 38]:

- The risk is great or the risk is small.
- The risk is high, significant, etc.

Since the assumed complex event as combinatorics of events that can cause possible damage is unlikely with respect to a probabilistic measure, *it is only significant that it is recognized* that a negative event will occur (may occur). It is obvious that in this case it is necessary to determine only the value of the supposed losses (damage).

However, in card games, some combinatorial events as chains of occurrence of "moves" can be determined very clearly. But their "frequencies" or even "probabilities" *are objectively fuzzy* due to the smallness of the possible frequency of the predicted combination among a multitude of other possible combinations. There are cases of lucky players who could calculate both combinations or scenarios of moves and even approximate frequencies that allow avoiding obvious losses in the game.

The following conclusions follow from the analysis:

(a) *In discrete states of the system* (in particular, in states with a game structure), it is almost always possible to *clearly identify moves, chains, scenarios* of events and calculate possible damages in specific chains of moves, but *without using probability values for results*, as these *probabilities (values) are small—"near-zero"*.
(b) Identification of all supposed moves is possible only because, for example, in card games *all elementary events are discrete, and a set* of scenarios, although large, is *countable* [29, 39].

This provision indicates the possibility of clear calculation of combinations of moves in card games with a set, for example, of 36 cards and with a combination of symbols on cards, including the form of signs (and even color), for example, "diamonds", "clubs", "hearts", and "spades". Theoretically, it is possible, *due to countability*, to predict the whole set of elementary sequences of moves; the only question is the complexity of such procedures, aside from using a computer.

In people's games, the ability to "calculate moves" depends on the intelligence of the players or on the perfection of the computer performance of the game procedures, which in principle is of no importance for assessing the "*risks*" and "*chances*". It is important to note that *all events* defined by each individual card *can be attributed a clear measure of the measurability of events* by their use, as is customary in the theory of probability spaces (i.e., in the probability theory) [29]. Something similar is characteristic of chess games, which also makes it possible to create chess (computer) machines or computer players.

However, in real highly reliable technical systems that differ from "card" systems, the measure of the possibility of functional failure is uncertain and small (*fuzzy*). This is due to the fact that it is not possible to *clearly describe the "tails of distributions"* (pdf) for RT opposite events (*such as "FAILURES"*), where critical damages can occur that *cannot be discarded* even if the *probabilities of events are "near-zero"*, for example, in ATC [11, 40, 41].

2.3.2 *Mathematical Basis of Risk Models as a "Risk Measure" (According to the RAS)*

The mathematical basis of the probability theory for hazardous (risk) events can be the axiomatic probability theory by A. Kolmogorov, the core components of which are countable sets of elementary discrete events, and a set of combinations is countable sets of discrete events from the σ-algebra. The rare events problem derives from these positions, but due to the physical characteristics of the quality indicator of "safety" type, and also because of the rarity of events, this problem should be studied separately, as announced at ICAO.

The probability space U can be defined as [29, 39]:

$$U = \left(\Omega, E, P | \sum_{0} \right), \tag{2.2}$$

Where Ω, E, P is a set of conditions of the game, rules for making moves, attributing properties such as consequences to events, etc.

Therefore, the following assumption can be accepted as substantiated: If the σ-algebra E determines only clear combinations of various events of any kind in U (e.g., events in a class of risks or chances), then the measure P, if it is small in this class, *is completely unnecessary in the "rare events problem"*.

This provision is decisive in the sphere of technical and economic systems and even suitable for natural complexes in assessing the possibility of random occurrence of events and damages. *Damages*, as is known, *if they are not average damages*, do not depend on the randomness measure, but depend specifically only on the properties (significance) of external factors and the design of the system with its reactions to the affecting factors.

The thing is that in technical systems the main task is to ensure a high value of efficiency from the standpoint of consumer properties.

From the standpoint of the reliability theory, this means providing, for example, a high level of efficiency (fail-safety) or success. Ensuring high *consumer properties requires obtaining a standard reliability indicator with a probability close to one* [10, 33, 34, 42].

The whole problem (the *main problem*) is that when assessing risks, the *significance of large losses* (damages) is taken into account *for events* (risk events) *opposite to the main event* (in quality). *Opposite events, although rare,* as in the games theory, *are unacceptable* for economic or other reasons.

This also applies to classes of tasks in assessing the safety of systems.

However, system safety theory (SST) *has nothing in common with the RT* regarding the technical equipment and procedures for assessing consumer properties.

The RT and SST (RRS) are different theories, although they are based on a single baseline; the SST (from the RRS) and the PSA are alternatives taking into account initial data, but there is nothing common in them regarding the equipment and indicators.

2.4 Mathematical Basis for the Definition of a Risk Event and an Integral Measure of Risk in the Probability Space

Substantiation of a New Definition of the "Risk" Category

Analysis of sources on the interpretation of the "risk" concept shows the following. The glossary of the American Oxford University (1998 and 2012 editions) states:

"Risk" is the possibility of the occurrence (with some measure) of severe (negative) consequences in the supposed situation under certain conditions (*such as threats*).

Thus, here the risk is not the "probability", because "probability" is a measure of randomness.

In the RAS publication (ICS—Malinetskiy et al. [9]), it is proved that it is most correctly and scientifically substantiated to adopt the following initial definition: "Risk is a measure of danger".

The concept of the "danger" definition has not yet been standardized (this, however, is taken as such below as a postulate).

The fundamental difference between mathematical objects in safety systems, as mentioned above, is that rare events are studied (events opposite to A, as in the RT) instead of events A (quality) considered in the reliability theory. This means that events are studied that are complementary to A in the binary outcome space:

$$\overline{A} = 1 - A(\text{logically}). \tag{2.3}$$

It can be assumed that in a technical system with normal quality indicators, which are guaranteed by the reliability theory methods, random events of the type "*non-failure*" ~ A are studied.

The quality of such systems in the RT is determined from the initial event A with consumer properties in the form of indicator P_A, for example, the "probability" of an event:

$$P_A = P\left\{A|\sum_0\right\}, \tag{2.4}$$

P_A is a *non-random measure of the measurability* of such a random event.

Physically and statistically, P_A denotes the quality of objects A in a certain set $\{A|\sum_0\}$ under conditions of existence for $\sum_0$.

Indicator P_A determines the measurability of a random event and is a non-random measure of the "*amount of randomness*" in the specified set. In highly reliable systems, these indicators are large in their values and close to one:

$$P_A \sim 1.$$

It must always be kept in mind that P_A is a real, non-random, clear number that can be found a priori analytically (under ideal or approximately ideal conditions) or even a posteriori for some reliable statistics from experiments.

"Safety" in the RT is assessed in the appropriate "state" by the "*level*" *of severity of the consequences* [10], but always *only for opposite events* $\overline{A}$, complementary to A, such as "*failures*".

In the RT, these events are incompatible and form a parent entity in the form of, for example, a binary outcome space Ω under a binary partition:

$$\Omega = A \cup \overline{A} \cup \varnothing(\text{incompatible } A \text{ and } \overline{A}) \tag{2.5}$$

where $\varnothing$ is an "empty" element.

Thus, both in the RT and in the SST, the objects that are the basis for assessing some properties of type (2.3) and (2.4) are different and opposite, always *incompatible* and specifying different non-coinciding properties.

Thus, both in the RT and in the SST, the main position is the same; namely, it is accepted that "*safety*" *is assessed through* "*danger*". For this purpose, the RT uses additional characteristics of the "*consequences from certain failures*" in the form of "criticality" of failures [6, 10, 26, 43]. Therefore, it is necessary to introduce the concept of a "hazardous" $A_* \equiv \overline{A}_*$ or "risk" event R, such that this event A_* entails negative consequences. Thus, A_* is not always a trivial failure $\overline{A}$ with weak consequences, but a critical one with the sign (*):

$$R \equiv \overline{A}_* = \overline{A}_*(\omega_{\xi*}|H_R), \tag{2.6}$$

where H_R is the designation of the value (magnitude) of the negative result in the form of some damage; $\omega_{\xi*}$ is an elementary random event with some signs of randomness ξ; $\overline{A}_*$ is a class of events.

The difference (divergence) between the RT and the SST positions in "safety issues" consists in the occurrence possibility for events of the $\overline{A}_*$ type.

2.5 PSA and SST Safety ("Hazard") and "Risk" Models

The PSA theoretical apparatus for assessing safety in the RT is the probability theory, "based mainly on a posteriori statistics". Therefore, relation (2.4) defines clear objects in the form of measurable random events A and $\overline{A}$ with clear measures of the property, namely the property of applicability of the methodology for calculating probabilities of events with the help of analytic functions like pdf, Prdf.

Therefore, the objective conditions for the "reliability theory" to come to a "dead end" in addressing the problem of ensuring safety with rare events were generated by the very same theory as it was refined in the field of achieving the required quality by standard indicators of reliability (*fail-safety*) of the systems.

Systems have become highly reliable under a set of certain conditions $\sum_0$. This turned out to be sufficient for the practical needs of engineering, industry, and customers.

But a contradiction arose between the RT and the SST: F*ailures became so rare that it was impossible to estimate them*, but the consequences of some (rare) failures turned out to be so significant that it was *unacceptable to ignore them* because of their probability [44].

From a mathematical point of view, the objects of the *reliability* system ("*properties*") and *safety* system ("*states*") models are different (*opposite to each other*).

The main RT objects are *events A characterizing the system operability* without special restrictions. Events A are measured clearly on the Boolean truth lattice by *clear indicators of the properties of objects—through probabilities* $P_A = P\{A|\sum_0\}$, and the probabilities are determined clearly (without errors)—analytically through Prdf and pdf (i.e., these functions should be known).

These *Prdf and pdf are true*, or *true with errors in confidence intervals*, which is measured by the randomness of the main event, as above; i.e., there is "truth" only for the main event A. Indeed, for $P_A \to 1$ this is completely unimportant, since the quality of the system is ensured through the property indicators (and through probabilities).

The SST objects (as in the RT) are also events, but risk ones $\overline{A}$, opposite to A; such that

$$\Omega = A \cup \overline{A} \cup \varnothing \Rightarrow \overline{A} \equiv \Omega \backslash A, \tag{2.7}$$

where sign (\) denotes subtraction and $\overline{A}$ complements Ω.

In (2.6), the outcome space Ω for A is contained clearly within the Boolean lattice through $\overline{A}$. However, it is known that in the RT only the *equivalent replacement of a clear* P_A *value by its fuzzy estimate* $\tilde{P} = P_A(m, n)$ works practically, i.e., $P_A \equiv \tilde{P} = \tilde{P}_A(m, n)$, where n is the total number of tests (trials) and m is the number of successes over A [10, 45].

This statement is accepted by default, since for practical purposes it is of no importance. Consequently, $P \equiv \tilde{P}(m, n)$ can be found on a Boolean lattice for particular situations and a probability space can be constructed with [35, 36] taken into account, as shown in (2.1).

That is why incorrect PSA estimates appear for $\tilde{P}_{\overline{A}}$ from (2.5), for example, in the work by B. Orlov [16], when the sign (+) is attributed to the measure of hazard (risk), as another incompatible event $\overline{A}$ complementary to A in (2.5) is ignored.

In this case, the errors $\Delta \tilde{P}_{\overline{A}}$ in determining the probability of an opposite event $\tilde{P}_{\bar{A}}$, with the analytical dependence $P(\overline{A}) = 1 - P(A)$ using $\tilde{P}\,(m, n)$, cannot be estimated due to the "tail" [9]. The establishment of the error rate $\Delta P(\overline{A}|n, m)$ for determining the probability $P(\overline{A})$ of an opposite event $\overline{A} = 1 - A$ (in the logical sense) does not make sense. This yields nothing to solve the problem of safety. In highly reliable systems, the parameter $m < n$, while it is obvious that $\Delta m = n - m$ is extremely small for rare events. Therefore, to improve the accuracy of estimating the indicator $\tilde{P}_{\bar{A}}$ by increasing $n \to \infty$ is inexpedient for economic reasons and does not make sense, since the main goal of quality assurance is only to achieve a high level of the specified value, which is usually achieved already at $n \sim (50–100)$. At the same time, the considered estimates are both consistent and effective [39, 46]:

$$P_A\left(\sum_0\right) \to \tilde{P}_A\left(\sum_0\right) \text{ through } \tilde{P}_A(\text{m,n}). \tag{2.8}$$

For example, in non-recoverable highly reliable systems, the measurable random event A has the form

$$A = \{\tau \geq t_*\} \sim A_\xi \sim \omega_\xi = (\tau > t_*),\ A = \left\{A(\xi)|\sum_0\right\}, \tag{2.9}$$

where τ is the random time to failure and t_* is a certain critical value of the operating non-failure time, for example, the standard average operating time to the first failure. This condition is approximately the same as for Cox and Smith renewal streams [33, 46].

Usually, according to the technical requirements, the following is given

$$P_A = P(A) = P\left\{\tau \geq t_* | \sum_0 \right\}, \tilde{P}_A(m,n) \sim P_A = P_A(t_*). \quad (2.10)$$

$$P_*(A) \sim P_A(t_*) \sim 0.95;\ 0.98;\ 0.99;\ 0.999;\ 0.9999,$$

where t_* is the quantile of the corresponding Prdf, and in (2.8) this quantile has a relatively small spread.

Small probability values $P_{\overline{A}}$ for rare events of the type $\overline{A}$ (i.e., "not A") opposite to the initial (main) event A such that $P_A = P(A) \sim 1$ are determined with large errors $\Delta P(\overline{A})$.

This means that minor errors $\Delta \tilde{P}_A$ in determining the main indicator $P(A) \sim 1$ that lie within confidence intervals are not essential for assessing the quality of systems with the probabilistic reliability indicator with $P(A) \sim 1$. However, $\Delta \tilde{P}_{\overline{A}}$ can be significant if probabilities for events $\overline{A}_*$ are determined in the "tails" of pdf, as it is shown in Table 2.1 on the example of the NASA material analysis [1].

2.6 Comparison of RT and SST Quality and Safety Indicators

2.6.1 *Estimation of Errors in the Experimental Determination of the Probability*

The effectiveness of methods of the classical reliability theory in *quality* (*not safety*) *systems* is due to the use of parametric indicators $P(A)$ for a random event A. In practice, P and the parametric empirical (a posteriori) evaluation $\tilde{P}(A)$ of type $\tilde{P}(m,n)$ are indistinguishable within the limits of confidence accuracy [10], where n is the number of tests for the object E to check the property $\tilde{P}(A) \cong P_*(A)$ through m and n; m is the number of outcomes favorable for A; $\Delta\tilde{P}(A)$ is an allowable confidence error for the estimate $\tilde{P}(A)$; ~ is the correspondence sign. Within the standards, the estimates obtained are effective and consistent [10, 15].

Indeed, the error $\Delta\tilde{P}(A)$ for the main event (for quality) is quite acceptable for a highly reliable system. But in this case, the error in determining $\Delta \tilde{P}_{\overline{A}}$ for the probability $P(\overline{A})$ of an opposite event $\overline{A}$ based on $P \equiv \tilde{P}(m,n)$ with the ratio $m(\overline{A}) = n - m \sim$ 1–2 cases to n turns out to be unacceptably large. But this has absolutely nothing to do with determining the operability (fail-safety) indicator $P(A)$.

Thus, for theoretical values of $P(A)_*$ from (2.1), the corresponding series of theoretical values of $\tilde{P}(A)_*$ is obtained. Thus, with the help of RT formula (2.8), the error $\Delta P(\overline{A})$ in determining the probability $P(\overline{A})$ of opposite events can be unacceptable; i.e., the percentage error is large. This determines the range of applicability of RT methods for assessing *safety levels through “consequences from functional failures”*—i.e., “safety” through the level of “hazard”, as shown in (2.3).

At the same time, the basic provision on safety in both the RT and the SST is the assessment of the admissibility for the consequences of “failures”.

That is, the main task here is the definition of a clearly opposite event $\overline{A} = \Omega/A$. Consequently, for the term “risk”, it is also necessary to find a clear measurability measure for any events through “probability”, i.e., for $\overline{A}$ also (but in highly reliable systems (according to the PSA), it is extremely difficult).

So, $\overline{A} \Rightarrow P(\overline{A})$ turns out to be immeasurable, since there is no a posteriori probability $P(\overline{A})$ as there is no necessary statistics for $\overline{A} = \overline{A}_* = R$; i.e., there is the statistics for $\overline{A}$ taking into account the property $H \neq 0$ (nonzero damage).

Therefore, for $\overline{A} \rightarrow \underset{\sim}{\overline{A}}$, some already *fuzzy randomness measurability measure for such an event* $\underset{\sim}{\overline{A}}$ must be found. This measure follows from the methodology of fuzzy risk calculation and is described in the SST. Since this fuzzy measure cannot be a “probability” due to the rarity of events and due to the prediction of “possible damages” with possible fuzzily measurable random events $\underset{\sim}{\overline{A}}$ and possible damage H, the following is assumed:

$$\underset{\sim}{\overline{A}} \sim (\overline{A} = \Omega/A) \Rightarrow (\underset{\sim}{\overline{A}}|H) \rightarrow \underset{\sim}{R} \sim R, \tag{2.11}$$

where $\underset{\sim}{R}$ is a fuzzy event (risk) and (/) denotes subtraction, as in (2.5).

This fits into the ICAO risk concept for predicting the possible consequences of a risk event, such as R or $\underset{\sim}{R}$:

$$\underset{\sim}{R} = \underset{\sim}{\overline{A}} = \underset{\sim}{\overline{A}}(H|\Sigma_0) \equiv \underset{\sim}{\overline{A}}. \tag{2.12}$$

On the basis of (2.10), for R, due to immeasurability of $\underset{\sim}{\overline{A}}$ and instead of the probability, a fuzzy randomness measure of type $\mu 1$ for the event $\underset{\sim}{R}$ is introduced and possible damage H is indicated, as was done in [20].

The SST risk concept given for the first time in [14] can be formalized in the form (2.11), where R is a clear (fixed physically) random event, and $\tilde{R}$ or $\underset{\sim}{\hat{R}}$ is “risk” as a fuzzy measure of the predicted amount of hazard in the system state with a predicted (possible) risk event R for a specific (clearly defined physically) predictable threat:

$$\underset{\sim}{\hat{\mathrm{R}}} = f_R(R|\Sigma_0) \sim \hat{\mathrm{R}} = f_R(\mu_1, H|\Sigma_0). \tag{2.13}$$

where in (2.13) the fuzzy measures have sign “wave” below.

The integral estimate of the “risk” $\underset{\sim}{\hat{\mathrm{R}}}$ as a fuzzy amount of danger in a given state can be expressed in the following *non-vectored* indicators:

Option 1—Linguistic variable as a predicate definition that can be well-defined or fuzzy (based on fuzzy sets);

Option 2—"Table diagram with a risk assessment matrix" (as shown in Fig. 5.1 in Chap. 5, see below).

An integral estimate $\hat{R}$ or $\underset{\sim}{\hat{R}}$ is either given by a "number" or by a combination of numbers as a "measure of randomness and severity of consequences" (e.g., in the form of a "Boeing matrix" in Fig. 5.2). It should be assumed that $\overline{A}$, R are events (measured physically clearly and taken from the universal set of characteristics and properties of the system and external effects); $\underset{\sim}{\overline{A}}$, $\underset{\sim}{R}$ are fuzzy events; $\tilde{R}$ is a risk situation estimate through two or three elements; $\underset{\sim}{\hat{R}}$ is integral fuzzy risk estimate (such as "higher", "lower"); f_R is an operator function for determining the integral level of risk (clearly by ICAO indicators or fuzzily in linguistic variables or points); ~ is the correspondence sign (the fuzziness sign indicated below).

The parameters in (2.11) *do not constitute a topological space; therefore, the presented relation is a "tuple"*.

The safety of systems can be ensured, within the framework of acceptable requirements only by reducing the risks of catastrophes and on the basis of risk and risk factor management methods in accordance with methods of the system safety theory [8, 19].

2.6.2 Two-Dimensional Estimate of Risk Significance of an "Amount of Hazard"

Based on recommendations regarding the interpretation of the ICAO risk concept [27], a full set of relationships is proposed here based on (2.11) for estimating the safety level by comparing the potential (calculated) risk $\tilde{R}$, $\hat{\tilde{R}}$ with the level of acceptable risk $\tilde{R}_*$, $\hat{\tilde{R}}_*$ through the projected consequences in (2.11) (damages $\tilde{H}_R$) and other indicators of the properties of rare hazardous (risk) events:

$$\tilde{R} = (\mu_1\, \tilde{H}_R\, |\, \Sigma_0), \tag{2.14}$$

$$\hat{\tilde{R}} = f(\tilde{R}\, |\, \Sigma_0) \tag{2.15}$$

where μ_1 is a measure of risk of the first kind, indicating the uncertainty (or randomness) of occurrence for a risk event R with a negative result $\tilde{H}_R$; $\tilde{H}_R$ is a measure of the consequences or damage (the cost of risk is the "severity" of damage); Σ_0 denotes experience conditions or parameters of the situation in the operation of the system, including the hazard class, the system hazard model, the structure of the RT failure tree, the state change graph, states of catastrophic system failures specified by the method of minimal cut sets of failures (according to Ryabinin [22], Makhutov [15]); $\hat{\tilde{R}}$ is an integral risk (with fuzzy estimates according to (2.11), i.e., the amount of hazard in a given state, as was assumed above.

Description of the set of conditions Σ_0 for the existence and definition of the system includes some of the characteristics indicated above.

The most important should be considered a "hazard model" according to the samples from Sect. 2.5 (see above).

The presented relations (2.14) and (2.15) reflect the methodology for determining the significance of risks and the application of the risk analysis matrix (by ICAO) based on the methodological provisions of the Fuzzy Sets (and subsets) theory.

Here, interpretations of the introduced concepts are given within the framework of concepts for Fuzzy Sets by M. Fujita (Tails—far from medium [11]) and G. Malinetskiy ("Hard Tails" from the Risk Theory—RAS ICS) [9].

The scientific problem is the construction of functions for estimating the quality of a set of elements in (2.12) and (2.13) that, in the general case, are not part of vector spaces that are dense everywhere.

In the general case (fuzzy estimates, etc.), it is inadmissible to normalize the set (2.12) to a scalar vector convolution, since its elements do not form any topological space, according to [33, 46], since it is impossible to find the vector norm, as described in [15, 47].

2.7 Decision-Making Regarding Risks and Chances in Monitoring and Ensuring Safety in Civil Aviation

If the 30-year-old "risk concept" ("*risk is a probability ...*") will be followed further, then the summation of risks should be accompanied by the construction of an outcome space, the distribution of probabilities by hypotheses of conditions, damage, etc. This is practically impossible if the probability of rare events is "near-zero".

However, it is possible to solve the *problem of making "optimal solutions" by "weighing the risks" and "chances"* on the basis of the RRS (SST) provisions.

In the new paradigm, within the framework of the *Fuzzy Sets*, it is possible, using the formulas (relations) presented below, to consider and to find, with a fuzzy measure of randomness for hazardous events (such as catastrophes), various options for determining the units of risk measurement.

For example, "clear risks" (in the sense of the specified values or levels of significance) can be presented in the form of ICAO–Boeing indicators:

- "Conditionally *clear* risks" in the form of known combinations of "two or three" elements (3E, 4D) according to the NASA diagram [1] from the adjusted matrix (these combinations are also strictly fuzzy);
- "Fuzzy risks" in the form of linguistic variables with fuzzy (undetermined) randomness and with fuzzy damage;
- "Risks as PSA probability"—with the appropriate statistics;
- "Fuzzy [40] and clear [11] risks according to Reich's formula in the form of the average number of possible (expected) aircraft collisions per flight hour", for

example, in ATC at the opposite flight levels with *TLS* indicators per ICAO (*M. Fujita—Tokio, Japan*) with "heavy tails";
- "Medium risks, as the average loss value" according to N. Makhutov [15] (although this indicator has no practical sense in the system safety theory *with fuzzy estimates of safety performance*, except in the case of estimating the average loss of aircraft in aviation units during combat operations). (*For example, it was stated above that the number of catastrophes for Airbus aircraft does not exceed 1–3 for 10 years of operation for approximately 10 million takeoffs and landings for the whole fleet of aircraft—old, new, and of various types.*)

This implies the need for a mathematical formalization of the ICAO concept:

ICAO Risk Concept: Likelihood & Severity of Harm.

It should also be noted that only the approach based on Fuzzy Sets made it possible to correctly introduce and substantiate the values $\hat{\tilde{R}}$ of risk assessment matrices adopted throughout the world. The scientific substantiation [8, 32] was first given by a group of authors (G. Gipich, V. Evdokimov, E. Kuklev, M. Smurov), which also enabled the creation of an adjusted ICAO matrix. This matrix was presented in Montreal and in Washington at the Boeing Company [48].

The value of the *Fuzzy Sets* method for assessing the safety of highly reliable technical (aviation) systems is quite important.

The fact is that many experts know the theory of *Fuzzy Sets*, especially its primary description in Russia in L. Zadeh's interpretation, but the use of the simplest applications of this theory in civil aviation was limited. The key point is to find relationships of L. Zadeh's theory with practice and with flight safety (FS). In *Fuzzy Sets*, there are key points that have not yet found wide application, for example, such concepts as a universal set of attributes—carriers of clear properties and their fuzzy subsets, etc., which is described in a separate RRS section—in procedures, methods, and algorithms.

As an example, here (and, perhaps, for future developments) the following tasks known in foreign practice and in civil aviation of Russia are formalized:

"***On three PICs and the departure decision under the conditions of wing and body icing***".

It is difficult to solve this task by the PSA method. Only one correct administrative decision is known substantiated by Boeing:

"Only take off when the aircraft has been treated and after the ice removal".

The task of "*finding the right solution*" is formed as follows. There are three independent aircraft and three independent PICs. However, by the ergodicity theorem, one can consider only a variant for one aircraft and one commander (PIC) who *selects one option from several alternative solutions:*

No. 1. *"Fly" by following the "hit or miss" principle without any "treatment"*, since *"Maybe nothing will happen"*, but the economy will be obvious, and the passengers will not be outraged, as the flight will not be delayed. However, *there is a risk obviously* that an accident may occur (and damage will be done).

No. 2. *"Do not take off until sunrise, as after the sun rises the ice melts"*. The risk of negative consequences is associated with losses due to the flight delay (but "What will the boss say?").
No. 3. *"Do not take off for a while, call special services, wait in line, treat the aircraft, take off following all standard procedures"* (according to Boeing). The risk of losses is manifested in the form of costs for the procedures and inconveniences for passengers, but there is a "chance" to survive in terms of the "icing" factor.

Initial conditions: departure events; icing events are rare; decisions of commanders are uncertain; there are no probabilities of events. It is *necessary to assess the "risks" and "chances"* and make the best decision following the PIC criteria (if the ATC intervenes, the solution will be different).

A rigorous solution to this problem using the PSA method is somewhat complex, but in the *Fuzzy Sets* method the solutions are more obvious and rigorous—*according to "Boeing"*.

2.8 Baseline of the RT to SST Transition with "Fuzzy Subsets" of RT Events Such as Functional Failures

The basis of this direction is the **analysis of the system safety as a state in a discrete probability space for *events with the "near-zero" probability***.

Initially, Fuzzy Sets of undesirable events are studied that cause negative consequences and certain damages in highly reliable systems. Probability (i.e., measurability of events) is not considered according to the RRS. Further, the methodology of the safety analysis is refined with the transition to the assessment of indicative risks (per ICAO).

This is done by *transitioning from the RT PSA to the SST in the area of fuzzy subsets* (see Chap. 3).

The main task arising from the presented outline is the definition and interpretation of the category of "*residual risk*" stemming from the RT concept and the *recalculation of this "risk" into catastrophes* that arise in highly reliable systems.

These problems were noted by the following authors: Aronov [10], Fujita [11].

The solution of these issues is discussed in Chap. 3 (Fuzzy Sets in the SST).

2.9 Possible Ways for Assessing the Safety Performance of Systems Based on the ICAO Methodology for Calculating Risks (Annex 19)

According to Annex 19, the current way is to find a solution to the problem of ensuring the system security based on new approaches that can avoid the difficulties of solving the problems of using the PSA for the case of rare events.

Many RT problems (theoretical and technical ones) have already been solved. A complete mathematical theory of the reliability of complex systems is created (Barlow-Proschan, B. Gnedenko, M. Kashtanov, E. Barzilovich, I. Ryabinin, I. Aronov, etc. [10, 33, 49, 50]). Mathematical fundamentals of risk assessment are developed for the class of probabilistic rally models.

Procedures and methodologies for calculating standard reliability indicators have been developed; standards, guides, regulations have been created; technical standards have been defined; maintenance of systems and facilities has been organized; systems for maintaining aircraft (functional) airworthiness in civil aviation have been established; the equipment for controlling and monitoring the state of products and system diagnostics (technical condition) has been refined [26, 30, 35, 42, 51]; principles of creation of highly reliable systems, assemblies, etc., have been developed.

Below in this section, the final evaluation of the comparison between the RT and the SST tools is presented. The limits of the applicability for the provisions of the classical reliability theory and the system safety theory are established that are intended to assist in investigations of catastrophic phenomena and determination of technical, economic, and social consequences of large-scale accidents at nuclear power plants, hydroelectric power plants, as well as aviation accidents and natural disasters.

However, unlike the Russian Federation, ICAO and the global community are gradually abandoning the PSA in the field of civil aviation and "space", and introducing the principles ("ideology") of "risks". But at the same time, the interpretations of this notion in theory are contradictory. However, significant practical methods have been obtained for using "risk analysis matrices" of risks that reflect the concept of fuzzy subsets, but this has not been realized as a theory (this trend is also typical for scientific work in the Ministry of Emergency Situations in the Russian Federation, and especially in JSC Russian Railways [52, 53]).

There are no works on the "safety theory" based on the idea of "fuzzy object analysis", characteristic for FO SMS systems.

From this, it follows that it is *advisable* to develop methods for assessing system safety by *switching to the theory of calculating non-trivial risks—the type of risks* caused by the possibility of the occurrence of rare events. The RT PSA method can only solve correctly problems of trivial risks, i.e., simple problems, not compound, for example, those that are not Reason chains or other.

This *transition* should be *based on the application of the fuzzy subset concept* and non-stochastic modeling of the *processes of changing the discrete states of systems in the form of J. Reason chains*, which are based on the MCF concept in the classical RT (Table 2.1 above).

It is important to note that all these positions were set forth in a number of works published in 2002–2007.

Suffice it to say that the following incorrect (non-scientific) phrases have migrated from ICAO to the RF civil aviation:

"*Risks for safety*";
"*Risks of hazard*" ("risk" IS the hazard);

"*Safe risks*", "unsafe actions" (verbatim translation of the FAA circular—2008), etc.

All this is incorrect, for example, in the opinion of the RAS Institute of Control Sciences [4, 9].

For example, it is necessary to use word as follows "*…risks of transmitting viruses* …", but one cannot say "*…risks of viruses* …", etc.

Considering this, some models of managed risks are being developed in problems of analyzing the possibilities of occurrence of catastrophes, for example, with spacecraft like the "Challenger" [15] and "Columbia" space shuttles.

In the SST solutions, probabilistic indicators are not used, only combinatorics of events is analyzed and paths to a catastrophe are identified.

Alternative methods for managing (and controlling) the system safety in the SST (RRS) are the following: *predictive (proactive and active) safety management methods (ICAO)* based on the logic of calculating the risks of occurrence of negative consequences detected in technical systems using SMS-like systems.

At the same time, the risk is *defined* as the "*amount of danger*" in given critical discrete states by fuzzy *subsets of object analyzed* in the formalized structures of the systems under study.

However, the probabilistic and statistical analysis of system safety (PSA) based on the use of RT methods has the following features:

(a) *The RT advantages are the completeness of the theory, the universality of methods and ways* for ensuring reliability ("recovery", "redundancy", "event tree", "failure tree"), and standardization of procedures based on large statistics.
(b) *Difficulties in the use of PSA in the RT* in safety ensuring and assessing arise when addressing risk management issues, in compensating for the instability of "statistics" for events with the "near-zero" probability, in the complexity of using the "Fuzzy Sets" apparatus, the discrepancy between the concepts and categories of the PF and FF in international and national standards for system "safety assessment". There is a known problem of "*launch on hunch*"—according to the PSA ([10]—NASA (Shuttle)).

Task 1. The so-called ***launch on hunch*** problem was revealed in the analysis of the peculiarities of occurrence of some catastrophes. The probabilistic risk concept stemming from the RT led, as it was found later [23], to erroneous solutions because of the inaccuracy in calculating the probability of a catastrophe for a number of design parameters, for example, for shuttle-type vehicles. After the loss of the "Challenger" [15], the "launch on hunch" concept was questioned in NASA [1]. But as it was explained, in the case of "Columbia", the concept of the approach considered earlier in NASA, took place, however, to a certain extent. Therefore, according to Annex 19, for developing theories of safety assessment in civil aviation, the concept of J. Reason chains of events and the calculation of risk indicators (risk levels) were adopted on the NASA recommendation without probabilistic indicators. This book shows that this confirmed the need to develop an RRS based on the Fuzzy Sets methods [see Chap. 3].

2.9.1 Area of Implementation and Standardization of the SST and RRS Provisions

Since the safety assessment methodology uses the "significance" of ICAO hazard indicators, through the "risk", some SST provisions should be adopted in the standards.

The following positions should be adopted in the standards using fuzzy measures of the properties of various mathematical objects of the systems under study.

- The system must be highly reliable so that functional hazardous events $R \sim R_*$ that determine the "residual risk", the residual fuzzy measure μ_R of randomness with a "near-zero" probability should also have PSA problems.
- The integral measure of risk $\hat{R}$ as the amount of hazard is a function (fuzzy or clear) of the set of 2 (or more) elements and the predicted measure of randomness (in any indicators, except the magnitude of the probability of a risk event).
- The measure of randomness of hazardous events is predicted mainly in fuzzy terms (and the damage H_R)—in linguistic, tabular forms—while it is unacceptable, for example, in JSC RR, to calculate some average expert probability of rare events, as recommended in RR standards.
- The integral risk (*fuzzy or clear risk measure*) as the amount of hazard can be shown also in points (as in the RAS) and within the modified risk classification by the RAS ICS (G. Malinetskiy).
- System safety is assessed by comparing the value of the calculated predicted risk with an acceptable level of risk in the predicted states with predicted and clearly identifiable threats.
- The "residual predicted risk" is defined through the fuzzy measure of the predicted hazard, for example, by applying quality systems that clearly give confidence limits for the probability of functional failure states in the range of probability of a failure event of 10^{-6}–10^{-3}, not worse.
- In the standards, it is necessary to adopt the concepts of "risks, threats, danger, safety" according to the concept of methods of predictable safety management (international conventional interpretation).

"*Risk is a possibility of occurrence of negative events* in the system taking into account the severity of the consequences from risk events" (i.e., "the risk" is not a probability).

In the regulatory documents, it is necessary to confirm that the indicators of the risk significance as a predictable amount of hazard with clearly defined physical threats are estimated by fuzzy indicators (higher, lower, high, etc.). The impact of threat factors with fuzzy parameters on the flight safety is also assessed fuzzily.

Thus, if we admit, *according to common sense, that "the probability of events is near-zero"*, i.e., the rare event under consideration is possible, but *Prdf and pdf for this event are unknown*, then the application of risk theory methods and fuzzy estimates of various safety indicators becomes substantiated.

Moreover, it is also possible to apply expert methods for risk assessment (as a method of predicting safety indicators without statistical material—and without the PSA).

The methodology derived from the ICAO provisions (from Annex 19) for calculating the risks of adverse consequences in aviation technical systems proved to be very constructive for the introduction of the system safety theory (SST) into the RRS.

In the SST, as described above, the basis for constructing hazard models is the *principle of risk assessment as a "hazard measure"* (*RAS*), or more accurately, in the RRS as a "measure of the amount of hazard" *without using indicators such as "event probability"*.

With this approach, the amount of hazard is measured as the amount of some single substance, for example, "water in buckets" with fuzzy amount ("by eye").

This provision allows for simple "summation of risks", "redistribution of risks", "risk taking", etc., "mitigation of consequences from risk factors", "avoidance of risk factors" (e.g., decide against landing to a "slippery runway" and perform a missing approach).

2.9.2 Methodical Recommendations on the Applicability of RRS Provisions in SMSs

Terminological Aspects. Regarding the ICAO approval of a new Annex (Annex 19) concerning the resolution of safety issues in aviation activities, it is advisable to follow some recommendations in civil aviation of the Russian Federation when developing SMSs (AA SMSs):

- The term "safety" to be interpreted by Annex 19 as "safety is a state …" and further "… through "risk". This term was adopted and included in ICAO Annex 19;
- To exclude the outdated definition "without hazard—a complex of measures … etc." [54, 55];
- To revise the concept "risk is … "a probability" of…" in the "rare events" problem, which will allow to avoid the need of calculation the "scalar—average risk" per Mushik-Muller [47] to assess the significance of "damage" in situations with hazardous events with the "near-zero probability" (this circumstance was discussed earlier in this chapter).

The safety indicator in the form of the "safe flight probability" $P_{SF}(\sum_0)$, which was widely used in civil aviation in the past per [54] for some complex of conditions Σ_0 introduced by (2.5), should be recognized, according to Annex 19, as insufficiently informative [56] as a characteristic of the "flight quality" for highly reliable systems.

Further, based on the results of the ICAO "hazard models" analysis discussed in this chapter, it can be argued that the known values from the world statistics of the arithmetic average indicators of negative events (aircraft destruction, passenger injury, etc.) *cannot be identified with the concept of "the probability of some events"*.

This can be explained by the fact that all such events are considered in total and, strictly according to the definitions [12, 39, 46], in different probabilistic spaces.

However, this characterizes best of all the significance of the "integral" *acceptable risk in any convenient* form.

2.9.3 On the Applicability of the NASA *(*ICAO*)* Formula for the Definition of *RMS Values for Random Variables*

It is known that in SMM [1] and [28] contains the RMS expression ~ σ of a random variable X:

$$\sigma = \sqrt{\frac{\sum_i x_i^2}{n} - m_x^2} \tag{2.16}$$

where n is the number of measurements x_i; $m_X$is the mathematical expectation of X. The practical significance of (2.14) is that the interval $\pm 3\,\delta$ is covered by a "bulb" of Prdf by 95% [11, 39]. In this case, it is possible to estimate the degree of variation of the X oscillations with respect to the average value. However, it should be noted that this relationship in solving problems with "rare events" should be used with caution. The fact is that m_X is not known in advance, and the number n is small (if catastrophes and accidents are concerned). Therefore, in the standards for SMS procedures, it may be necessary to determine the applicability of this relationship, for example, when analyzing statistics on incidents, etc.

The presented condition is valid in the general case only for central and absolute moments of a second-order distribution [39, 46] as follows:

$$DX = E(X - EX)^2 = EX^2 - (EX)^2, \tag{2.17}$$

where E is the designation of the process of finding the mathematical expectation, respectively, for continuous and discrete quantities:

$$EX = m_x = \int_{-\infty}^{\infty} x \cdot f(x)\mathrm{d}x \mathrm{or} EX = \sum_{i=1} x_i P_i. \tag{2.18}$$

A simple form of the quantities under consideration is obtained for Prdf according to the normal law or for the "bulb": $DX = \sigma^2$, but with a known pdf $f(x)$.

Thus (2.16) *is the very mistake in the theory of SMS, based on the postulates of rare events.*

2.10 Conclusions

1. The positions of the classical RT and the system safety theory (SST) regarding approaches for assessing numerical values of operational safety levels in the presence of functional failures in systems coincide incompletely, as methods of assessing the risks of negative effects in systems under the influence of hazards are different.
2. Differences arise from different interpretations of risk concepts: either within the PSA, through the probability, as in RT, or as in the SST—through the definition of risk as a measure of the amount of hazard.
3. The most important task of the methodology for assessing system safety levels is to solve the rare events problem, which is difficult to do correctly with the PSA-RT method.
4. Considered in the RT concepts of criticality significance for failures, including the concept of functional failure, differ in that the criticality in the SST is estimated without using the "event probability" and only with the measure of the event possibility (with the "near-zero" probability) and the magnitude of the projected damage in comparison with acceptable damage.
5. The transition to the new RRS doctrine makes it possible to "legitimize" the application of expert methods for assessing safety and proactive (predictive) management for risks of negative consequences in aviation activities (perhaps in other sectors, too). The essence of this approach is that the new concept of "risk" can be mathematically introduced into the system safety theory. Then, the incorrectness of some of the PSA procedures that occurs when investigating the properties of rare events is eliminated within certain limits.

References

1. Probabilistic Risk Assessment Procedures for NASA Managers and Practitioners, Office of Safety and Mission Assurance NASA. Washington, DC 20546, August. 2002 (Version 2/2).
2. McCarthy J (1999) U.S. Naval Research Laboratory; Schwartz N., AT & T. Modeling risk with the flight operations risk assessment system (FORAS). ICAO Conference in Rio de Janeiro, Brasil
3. Documents of the 37th ICAO Assembly (Oct 2011, Montreal).
4. Severtsev NA (ed) (2008) Kuklev E.A. "Fundamentals of the system safety theory", RAS Dorodnitsin CC, Moscow, pp 175–180 (in Russian)
5. Issues of the system safety and stability theory. / Issue 7. Ed. by Severtsev N.A. / Moscow: RAS Dorodnitsin CC, 2005. (in Russian)
6. Accident Prevention Manual. Doc. 9422-AN/923 (1984) International Civil Aviation Organization
7. Risk management—Principles and guidelines. Standards (Australia)—AS/NSZ ISO 31000 (2009)

8. Smurov MY, Kuklev EA, Evdokimov VG, Gipich GN (2012) Safety of civil aircraft flights taking into account the risks of negative events. J Transp Russ Fed 1(38):54–58 (in Russian)
9. Malinetskiy GG, Kulba VV, Kosyachenko SA, Shnirman MG et al (2000) Risk management. Risk. Sustainable development. Synergetics. Moscow: Nauka, 431p. Series "Cybernetics", RAS (in Russian)
10. Aronov IZ et al (2009) Reliability and safety of technical systems, Moscow (in Russian)
11. Fujita M (2009) Frequency of rare event occurrences (ICAO collision risk model for Separation minima). RVSM. ICAO, Doc. 2458. EIWAC, Tokio
12. Henley EJ, Kumamoto H (1984) Reliability engineering and risk assessment. Mechanical Engineering, Moscow (translated from English: VS Syromyatnikov (ed)). (in Russian)
13. Kuklev EA (1999) Predicting the occurrence of aviation accidents on the basis of chains of random events: collection of materials of the International Symposium MAKS-99 (CIA). (in Russian)
14. Aronov IZ (1998) Modern problems of safety of technical systems and risk analysis. Standards and quality, no 3. (in Russian)
15. Volodin VV (ed) 1993(Reliability in technology. Scientific-technical, economic and legal aspects of reliability. Blagonravov Mechanical Engineering Research Institute, ISTC "Reliability of Machines" RAS, Moscow, pp 119–123. (in Russian)
16. Orlov AI, Pugach OV (2011) Approaches to the general theory of risk. RFBR Grant-2010. Bauman MSTU. Moscow. (in Russian)
17. Orlovskiy SA (1981) Problems of decision making with fuzzy source information. "Science" FM, Moscow. (in Russian)
18. Rybin VV (2007) Fundamentals of the fuzzy sets theory and fuzzy logic. Study guide. STU Moscow State Aviation Institute, Moscow, 95 p. (in Russian)
19. British Standart (1992) Quality management and quality-assurance. Vocabulary. VS EN ISO-8402
20. Amer MY (2012) 10 Things you should know about safety management systems (SMS). SM ICG, Washington
21. Arnold VI (1995) Catastrophe theory. "Science" FM, Moscow. (in Russian)
22. Ryabinin IA (1997) Reliability, survivability and safety of ship electric power systems. Kuznetsov Naval Academy, St. Petersburg. (in Russian)
23. Waitmann M (2011) UN: Japanese nuclear power plants are not ready for a tsunami. www.metronews.ru, 02 June 11. (in Russian)
24. Annex 19: C-WP/13935—ANC Report (March 2013), based on AN-WP/8680 (Find) Review of the Air Navigation Commission, Montreal, Canada
25. Livanov VD, Novozhilov GV, Neymark MS (2013) Flight safety management system. IL SMS. "AviaSoyuz", No 1, pp 14–21. (in Russian)
26. Novozhilov AB, Neymark MS, Cesarskiy LG (2003) Flight safety (Concept and technology). Mechanical engineering, Moscow, 140 p. (in Russian)
27. SMS & B-RSA (2008): "Boeing", 2012.
28. SMM (Safety Management Manual): Doc 9859_AN474—Doc FAA (2012)
29. Kuklev EA (2005) Estimation of catastrophe risks in highly reliable systems. Materials of the 13th international conference "Problems of complex system safety management". RAS ICS, Moscow, 55 p. (in Russian)
30. CATS (Casual Aviation Technical Systems) (2012) Simulation of cause-effect relations in aviation systems on the basis of risk assessment. Research of the Air Accident Investigation Commission (Netherlands), ICAO (in Russian)
31. Bykov AA, Demin VF, Shevelyov YV (1989) Development of the basics of risk analysis and safety management. Collection of scientific papers of the Kurchatov Institute of Atomic Energy. Publ. House IAE, Moscow. (in Russian)

32. Smurov MY, Kuklev EA, Evdokimov VG, Gipich GN (2012) Development of tools for assessing the risks of occurrence of risks of AUI in the AASS of the airport system. J Transp Russ Fed 2(39). (in Russian)
33. Barzilovich EY, Kashtanov VA, Kovalenko IN (1971) On minimax criteria in reliability problems. ASUSSR Bull. Ser "Technical Cybernetics" 3:87–98. (in Russian)
34. Gipich GN (2005) The concept and models of predicting and mitigating risks in ensuring the airworthiness of civil aircraft. Moscow State University, TEIS, Moscow. (in Russian)
35. Itskovich AA (2010) Formation of requirements for ATS unit diagnostic systems based on the results of logic-probabilistic modeling. MSTU CA Scientific Bulletin, series PA LA, No 8, 18 p. (in Russian)
36. Kuklev EA (2005) Decision making in systems based on management of possible risks of undesirable consequences. Materials of the 13th international conference "Problems of complex system safety management". RAS ICS, Moscow, 87 p. (in Russian)
37. Kirpichev IG (2011) On the prospects and problems in developing the infrastructure for the maintenance of An-140, A-148 aircraft. Aircraft construction. "Aviation Industry", no 2, Moscow, 55 p. (in Russian)
38. Kabanov SA (1997) System management on predictive models. St. Petersburg University. St. Petersburg. (in Russian)
39. Korolev VY, Bening VE, Shorgin SY (2011) Mathematical foundations of the theory of risk. Fizmatlit, Moscow. (in Russian)
40. Dalinger YM, Isaev SA, Kuklev EA (2012) Risk management in the safety management system for RVSM. J Transp Russ Fed 6(43), 55 p. (in Russian)
41. Evdokimov VG, Martynov VV (2007) To the issue of system safety. Proceedings of the RAS ISA. Dyn Heterogenous Syst 29(1):116–120. (in Russian)
42. Bolotin VV (1998) The theory of reliability of machines, Mechanical engineering. Encyclopedia. Ed. board: K.V. Frolov (chair) et al. Reliability of machines, vol IV.-3. Mechanical Engineering, Moscow. (in Russian)
43. Accidents and catastrophes. Prevention and mitigation of consequences. Study Guide. In 3 books. In: Kochetkov KE, Kotlyarevsky VA, Zabegayev AV. Publishing House ASV, Moscow. (in Russian)
44. Documents of the ICAO High-level Conference (SMM). March 2010, Montreal.
45. Shor YB, Kuzmin FI (1968) Tables for reliability analysis and control. SOV Radio, Moscow. (in Russian)
46. Prokhorov PZ, Rozanov YA (1987) Probability theory (basic concepts, limit theorems, random processes). "Science"—FM, Moscow. (in Russian)
47. Mushik E, Muller P (1990) Methods of making technical decisions: Translated from German. Moscow: Mir. (in Russian)
48. Gipich G, Evdokimov V, Kuklev E, Mirzayanov F (2013) Tools: identification of hazard & assessment of risk. Report at the ICAO meeting "Face to Face". "Boeing" Corporation, Washington
49. Abramov BA, Gipich GN, Evdokimov VG, Chinyuchin YM (2013) Determination of the compliance service providers' safety management system to air transport standards. Sci Bull MSTU CA. 187(1):36–41. (in Russian)
50. Putin VV Russian nuclear power plants are reliable. (in Russian)
51. GOST 23743-88. Aircraft. Nomenclature of the flight safety, reliability, testability and maintainability indices. (in Russian)
52. General rules for risk assessment and management (resource management at life cycle stages, risk and reliability analysis management—URRAN). JSC Russian Railways standard STO RR 1.02.034-2010. Moscow (2010). (in Russian)
53. Procedure for determining the acceptable level of risk (URRAN). JSC Russian Railways standard STO RR 1.02.035-2010. Moscow (2010). (in Russian)

54. Sakach RV, Zubkov BV, Davidenko MF et al (1989) (edited by Sakach RV) Flight safety. Textbook. Transport. Moscow, 239 p. (in Russian)
55. Zubkov BV, Prozorov SE (2013) Flight safety. Textbook. Ministry of Education AMA. Ulyanovsk. (in Russian)
56. Evdokimov VG Integrated safety management system for aviation activities based on ICAO standards and recommended practices. J TR 2(45):54–57. (in Russian)

Chapter 3
Solving the Rare Events Problem with the Fuzzy Sets Method

The objective of this chapter is to create a *universal unified approach to assessing safety of complex systems through the ICAO risk concept*, using the tools of the Fuzzy Sets method. The idea of the Fuzzy Sets approach is set forth in well-known publications [1–4]. The problem is that the *concept of "risk"*, as was shown in Chap. 2, can be interpreted in the RSS as some *integral characteristic or measure of hazard*. At the same time, *using the value (level, cost) of the "risk", one can measure* the necessary *safety or hazard indicators* through some other indicators of the lower system level. At the same time, it is necessary to find ways of measuring risk (as a measure of danger) without probabilistic indicators.

The task in general (at the level of traditional methods) was solved by NASA [5]. The basis of the method was the techniques and algorithms of proactive risk (and system state) management, taking into account a set of risk factors and a risk assessment matrix. The theoretical basis of this NASA method is defined, by default, in Fuzzy Sets. But this was not formulated; perhaps, that is why the NASA algorithms demonstrated some incorrectness in the use of non-equivalent (incompatible) concepts. From the Fuzzy Sets perspective, this is interpreted as a violation of the binary truth validation rule (meaning "there is no third option") [2]. The bottom line is that in the risk assessment matrix, the concepts of randomness are applied equally: clear ones in the form of "probability" and fuzzy ones in the form of a linguistic variable. In the working materials of NASA (ibid., in [5]), the definition of "likelihood" (realistic possibility) is given (OXFORD Glossary), and the "probability" is officially indicated in the column, although fuzzy variables have been already adopted and recommended for use.

3.1 Axiomatics of Risk Models

The generally accepted definitions of risk types such as individual, social, political, financial, environmental, technology-related, constructional using a unified approach

Kuklev E.A. et al., *Aviation System Risks and Safety*, Springer Aerospace Technology, https://doi.org/10.1007/978-981-13-8122-5_3

designate only some features of the risk concept applicability when assessing the potential for hazard manifestation in the systems.

3.1.1 Principle of a Fuzzy Implication in the Analysis of Fuzzy Statements

The vagueness of statements or wordings results from the uncertainty of the content of concepts or descriptions for various reasons, including the lack of information about any subject or phenomenon. For example, in the method of confidence intervals (Chap. 1 of this book) not only is it impossible to predict specific values of the measured quantities, but even there is a duality in the assignment of the boundaries of intervals. This is especially true for estimating the probability values for the occurrence of hazardous events, which are determined in the areas of pdf "tails", which was discussed in detail in Sects. 2.1–2.4. The question that arises is to verify the truth of certain conclusions, for example, the significance of the integral level of risk in fuzzy terms: "higher", "lower", etc.

The solution of these problems is presented in [1–4] in the class of fuzzy implications of a set of fuzzy statements in an arbitrary set V. In this case, V is represented as a tuple of 2 parts: P denotes conditions (initial statements), and Q denotes results, but for selected statements $Q < V$, the truth of which is established with some measure μ with respect to P. In fuzzy logic (with fuzzy implications), the set P is not a cause, and Q is not a consequence, in contrast to implications in a clear logic. In fuzzy implications ($P \supset Q$), the elements of P and Q are selected quite arbitrarily, and the truth of the statements is checked. The set V is given as:

$$V = \langle P,Q | P \supset Q \rangle.$$

This set is mapped by the operator T in the interval [0,1] (according to L. Zadeh), and the result of the truth check is given by the formula under the selected criteria (Zadeh, Gödel, etc.):

$$T : V \to [0,1] \Rightarrow T(P \supset Q) \to [0,1).$$

The simplest (classical—according to L. Zadeh) fuzzy implication from V is a solution [2]:

$$T(P \supset Q) = max\{min\{T(P),T(Q)\}, 1 - T(P)\}. \tag{3.1}$$

Formulas of the form (3.1) are fitted or synthesized taking into account the specific features of applied problems.

In its content, the classical fuzzy implication (meta-implication)—unlike clear logic [4]—denotes the result of a binary logical operation in the form of a fuzzy statement with some measure of truth ($P \subset Q$).

Thus, in the NASA matrices [5], first *measurable clear probabilities* P_i (e.g., as shown in Table 2.4, by I. Aronov, Chap. 1) were replaced by *fuzzy linguistic variables*, and then the damage value was also introduced as fuzzy damage. The fuzzy implication resulted in elements (*binary tuples*) in matrix cells. The "safe corridors" of the matrix (type 5.1, see Chap. 5), by default, were found by the formula of fuzzy implication (3.1). But this was not indicated in the risk assessment matrices in [6].

It can be noted that it is incorrect to specify clear values (probabilities) and "work" with fuzzy linguistic variables: Only one type must be specified. (But in JSC RR [7], this is not specified.) *Note by the authors of the book*: When the aircraft is landing (when the glide path is captured), after receiving the information from the ATC controller, the aircraft commander can pronounce the phrase: "Damn slippery runway, landing can be dangerous". This is classified as *l-term* in the set *P* from (3.1). This term can be associated with some fuzzy statement *Q* containing the variant of the "solution". In this case, *P* can contain one more *P*-statement (condition) in the form of conjunction, for example, "Low visibility, the crosswind on the runway is significant". For the PIC, in terms of the common sense, the main task is to make a decision "Not to land and go around".

A fuzzy assessment of the degree of truth for these decisions will be: It is a "high risk" or "it is at least some chance"; i.e., this is "reality". In such problems, information on the values of the "event probabilities" does not yield anything useful, and procedures of type (3.1) indicate a method of theoretical investigation of similar problems within the "Fuzzy Sets" method.

In the ICAO and Boeing documents, instructions for the PIC actions are formulated in *clear logical implications*, and the predicted "fuzzy behavior" is compensated, to the extent possible, on the basis of the proactive and predictive state control of the "Aircraft-PIC" system using the NASA algorithm (4.8) from Chap. 4. The transitions to risk models (in Annex 19) described in Chap. 2 in the RRS doctrine (and in the SST tools) made it possible to offer solutions to some issues of the "rare events" problem.

3.1.2 Formula and Definition of Risk Significance

In this subsection, the physical interpretations of the risk concepts from Chap. 2 are formalized within the framework of the theory of discrete states. The uniformity of interpretations is achieved by studying hazardous phenomena in systems based on this description.

It is assumed that the risk $\tilde{R}$ as a mathematical category is the *predicted measure of the amount of hazard* (predicted hazard level by ICAO) for a possible (predictable) discrete event *R* (*present or absent judging from its manifestation*); i.e., $\tilde{R}$ is *improbability*. The discrete event *R* has *dual properties* [5, 8–10]. The estimate of $\tilde{R}$, i.e., the magnitude or risk of the occurrence of a predictive hazard event *R*, is initially always specified in a two-dimensional (or three-dimensional) set of indicators and in special cases is specified in an *integral form*, for example, according to the RAS classification [11] or using risk analysis matrices [12, 13], as noted in Chap. 2.

The most important in the proposed method is the identification and analysis of the possible conditions for the occurrence of "*catastrophes*" *as rare events* (according to ICAO), which have a measure of occurrence randomness of the "near-zero" probability. Therefore, it is necessary to transit to the application of the "*fuzzy subsets*" method instead of the probability theory methods.

In this connection, the theoretical RT provisions are initially considered here that form the essence of the PSA and are the cause of some insurmountable limitations of the PSA use in evaluating safety of the system functioning, taking into account the properties of rare events.

As is known, the *essence of the RT is a "hypercube of truth"* ("Boolean lattice") and some hypotheses. This implies the well-known systemic problems of the PSA use that can be avoided with the transition to the introduction of the SST and RRS doctrine provisions when solving issues of assessing the level of safety in case of rare events through "risks" in the new interpretation of Chap. 2. In this chapter, these provisions are considered in more detail.

Corollary 1 *The risk* $\tilde{R}$ *(value) as a physical category or its estimate should be evaluated through a two-dimensional or three-dimensional set of indicators, which can be written using two formulas similar to* (2.4) *from Chap.* 2. *Now, a three-dimensional estimate is additionally introduced*:

$$\tilde{R} = \left(\mu_1, \tilde{H}_R | \Sigma_0\right),$$

Or

$$\tilde{R} = \left(\mu_1, \mu_2 \; \tilde{H}_R | \Sigma_0\right), \tag{3.2}$$

where μ_1 *is a measure of risk of the first kind (uncertainty of the occurrence of a negative result—the degree of risk);* $\tilde{H}_R$ *is a measure of the consequences or damage (the cost of risk);* μ_2 *in* (3.2) *is a measure of risk of the second kind in the states manifested as factors, due to system errors and a threat to the state of the system;* Σ_0 *denotes the same experimental conditions or situation in the operation of the system* [*hazard class or system model, etc., from formulas* (2.2)].

The uncertainty of the occurrence of risk results depends on the methods of definition of the system S, *the state of the external environment, and the experimental conditions at the time when the corresponding chains of events* L_{k*} *begin to unfold at the time* t= t_0 *from the initial state* q_0 *of the system in the group of risk situations as*:

$$\Sigma_0 = \{e_k | q_0, t_0, S, \Gamma, g(\alpha, \beta)\}, \tag{3.3}$$

where $e_k = e_k(\gamma)$, $k = 1,2,\ldots$ are the characteristics of the studied structurally complex system S, such as the sign of the method to test the system π_S, its elements, the way of interconnecting the elements, the state of the environment and the presence of a terrorist threat [14], *signs of critical states* [15, 16] *in the system, a set of hypotheses* $\{g_i\} \Rightarrow R_1(g_1), R_2(g_2), \ldots$, *etc.*

When studying the same type (type 2) of catastrophes, $\tilde{H}_R = H_R = \text{const}$, *in other problems (according to the PSA),* $\tilde{H}_R \equiv \hat{H}_R$ *is the average risk (scalar), with the estimate* $\tilde{R} \to \hat{R}$ *"moving"* ($\to$) *to a scalar, but only in special cases when there is reliable statistics. Using a scalar* $\hat{H}_R$ *with the "near-zero" probability of event (risk) is inadmissible, except for special cases, as was shown in Chap. 2.*

The measure of uncertainty μ_1 *is a practical expert indicator of the* degree of risk (*higher, lower, or a simple quantity—as frequency from the event statistics or from the risk matrices, and in the case of a priori estimates—as the probability of the event). The constant value of damage* $\tilde{H}_R = H_R = \text{const}$ *taken in* (3.1) *or* (3.2) *in the study of catastrophic risks with estimates* $\hat{R}_*$ *indicates the possibility of the loss of a system of the same predetermined type for various causative factors and under various scenarios for the development of events leading to a catastrophe.*

Corollary 2 *In hazardous situations with the results of the "near-zero" probability, it is admissible to estimate the risk by relative and conditional indicators, as recommended by ICAO, and, at the limit, by the magnitude of the possible damage. This is the practice in civil aviation per ICAO, also used in property insurance or in* assessing the consequences of earthquakes.

3.2 Application of the Concept of Probability Spaces of the System Safety Theory in Fuzzy Risk Models

Within the framework of A. Kolmogorov's axiomatics [17], it is possible to separate [3] procedures of finding the structure of complex events and procedures for estimating the probability of events $P(\omega_i)$ or $A(\omega_i)$ in sigma-algebra $E \subset U$ of the probability space U [17, 18] by means of isolated operations with only elementary events $\omega_i \in \Omega$ from the outcome space Ω.

In the proposed method, the risk events R of various nature [7, 19, 20], as shown above, are interpreted as random discrete events with dual properties in the form of randomness and mandatory manifestation of negative consequences in the form of a certain damage. Such events can be correctly described in the framework of the axiomatic provisions of A. Kolmogorov's probability theory [17, 18], including the use of the sigma-algebra E of elementary discrete events—outcomes $\omega_i \in \Omega$, $A(\omega_i) \in E$ in the probability space U:

$$A(\omega_i) = \cup\, \omega_i;$$
$$\bigcup_k A(\omega_i) \bigcap_j A(\omega_i) \subset E,$$

$$U = (\Omega, E, P). \tag{3.4}$$

The subset E of U in (3.4) can also be introduced in the form of a Borel sigma-algebra under the quadrature of continuous spaces X_i in the case of a parametric dependence of the measure of discrete event randomness $c\mu(\omega_i)$, $\mu(A(\omega_i))$ on some continuous argument $\tau \sim (\tau \in [0,t])$. This argument denotes a random moment of time the selected event occurs in the time interval $A(\omega_i)$ of observation $\tau \in [0,t]$: $\cup,\cap$ are, respectively, the union and the intersection of events.

The following provisions are accepted.

- A discrete elementary event ω_i or class $A(\omega_i(j)|q_j) \in \Omega \subset E$ is defined as a result of the change $q_{j1} \rightarrow q_{j2}$ of some discrete states of the system $q_i \in Q$ that are points in the hyperspace Q formed by the Cartesian product of discrete spaces X_i characterizing the properties and structure of the system under study:

$$Q = X_1 \times X_2 \times \cdots X_i \cdots \times X_n, \tag{3.5}$$

$$q_j \Rightarrow A(\omega_i) \equiv A(\omega_i(j)|q_j) \subset E, \tag{3.6}$$

the sign $\Longrightarrow$ means "results in", and for random events $R(\cup\omega_i) \equiv A_*(\{\omega_i\}),\ A(\omega_i) \subset E$, * is sign of the criticality of consequences in the class $R \in E$;

- A risk event A_* is a class of events $A_* \equiv R = \cup R_j$ composed of incompatible private risk events R_j, i.e., from alternative events (ways) of the system ending up in a catastrophic state of a given type in a subset of states $Q_* \subset Q$ of the system.

The ***axiomatic definition of a risk event R causing damage*** is introduced in a sufficiently proper way by mapping outcomes ω and events $A(\omega) \in E$ from (3.4) to the phase Borel probability space $U_B \Leftrightarrow U$. Here, $\forall\omega \in \Omega\, \exists\xi = \xi(\omega) \in U_B$ with the distribution P from (3.4), where ξ is an arbitrary function of ω in the form of a real quantity. If $\xi(\omega)$ is given a physical meaning and dimensionality of the damage magnitude $\tilde{H}$ for each $\omega \in \Omega$, i.e., $\xi(\omega) = \tilde{H}(\omega) = H_\omega$, then event $A(\omega)$ in the sigma-algebra from (3.4) takes the form:

$$A(\omega) \rightarrow A_*(\omega) \equiv A_*(\omega, H(\omega)) = R(\{\omega\}|P(\omega), \Sigma_0, H(\omega)). \tag{3.7}$$

3.3 Assessing Significance of Risks in a Probability Space

The principle is based on the concept of a universal set with clear elements in the form of discrete states characterizing the occurrence of a set of discrete events studied in 3.4.

The system safety in the presence of a threat is determined, as was suggested above, through the "risk", in accordance with international standards [5, 12, 21], in the form of a system state with an acceptable (allowable) risk value $\tilde{R}_*$ for a risk event R, i.e., with the condition $\tilde{R} \le \tilde{R}_*$ according to (3.2), (3.3).

The final estimate is given in the form of an integral (generalized) indicator $\tilde{R}$ with a critical value $\tilde{R}_*$ found also on the basis of (3.1), (3.2) using the formula below:

$$\tilde{R} \to \tilde{R}_* \Rightarrow \tilde{R}_* = \left\{\tilde{R}_{*j}\right\},$$
$$\hat{R} = f_R\left(\tilde{R}_*|\Sigma_0\right) \equiv f_R\left(\mu_{1*}, \mu_{2*}, \tilde{H}_*|\Sigma_0\right), \tag{3.8}$$

where f_R is the functional that has, in particular, the form of the operation of scalar convolution of the set of elements from $\tilde{R}$ in (3.1) or (3.2) for the μ_1-averaged damage estimate $\tilde{H} = \tilde{H}(\omega_i) \sim \tilde{H}(A(\omega_i))$ as the average damage $\hat{H}_{R*} \equiv \hat{R}_*$ over all μ_1 from the corresponding distribution in the probability and $\hat{R}_* = \hat{R}_*(\tilde{R}_*)$, $\hat{R}_{**} = \hat{R}_{**}(\tilde{R}_{**})$.

The integral value (level) of risk $\hat{R}$ in (3.8) can also be found by means of operations of assigning preferences to various combinations of elements in sets $\hat{R} \to \hat{R}(\tilde{R})$ in formulas (3.1), (3.2).

At the same time, the significance of the risk $\hat{R}$ or its estimates $\tilde{R}$ for the event R is determined by comparing the actual (calculated $\hat{R} < \tilde{R}_{**}$) risk (3.8) with some of its critical values $\tilde{R}_*, \hat{R}_*$ or $\tilde{R}_{**}, \hat{R}_{**}$: $\{\tilde{R} < \hat{R}_*$: or$\}$ are risks R that are admissible (acceptable) by estimates $\tilde{R}$; $\{\tilde{R}_* \le \tilde{R}_{**}$ or, the maximum permissible risks, $\tilde{H}_{**}$ or $\hat{H}_{**}$ are integral maximum permissible values of possible critical (admissible) damage. By space (1); (*) is the sign of the indicator values criticality, respectively, $\hat{R} \to \hat{R}_*$, $\tilde{R}_* \to \tilde{R}_{**}$, as compared with the values $\hat{R}_{**} \sim \tilde{R}_{**}$ of the admissible critical risk with the index (**), so that $\hat{R}_* \le \hat{R}_{**}, \tilde{R}_* \le \tilde{R}_{**}$, analogy with (3.8), the following can be written:

$$\hat{R}_{**} = f_R\left(\mu_{1**}, \mu_{2**}, \tilde{H}_{**}|\Sigma_0\right), \tag{3.9}$$

then $\hat{R}_* = \hat{R}_*(\tilde{R}_*)$, $\hat{R}_{**} = \hat{R}_{**}(\tilde{R}_{**})$.

In the case of catastrophic risks, as mentioned above, $\tilde{H}_R = H_R = \text{const}$, since the concept of average risk $\tilde{H}_R$ in (3.9) or $\hat{R}$ in (2.7) (from Chap. 2) in this case has no practical meaning (is absurd, according to [18]).

3.4 Interpretations of Fuzziness for Subsets of Factors in Risk Analysis Procedures Based on ICAO Recommendations (from SMM-Doc 9859)

3.4.1 Effects of Pdf Fuzziness on Risk Indicators

The decision on the need to take into account the fuzziness of the values of the risk factor subsets is determinative in the theory of SMS construction (per ICAO). This is due to the fact that ICAO has put forward the concept of implementing safety management through proactive control for activities of airlines or subdivisions of other level using methods for calculating the risks of undesirable consequences when certain (rare) random events occur, which is accepted in [5, 22] by default, but is given in the SMM text [6] without any substantiation.

The transition to this concept has led to contradictory interpretations of the significance of risk events identified proactively in predicted situations characterizing the aircraft operation.

ICAO's primary positions are based on the reliability theory methods (as in the case of UAS) that make it possible to determine safety indicators with confidence up to $\alpha = 0.95–0.98$ (confidence interval) with acceptable error level of $\varepsilon = 0.05–0.1$. This interval usually determines the characteristics of failure flows in aircraft systems, at nuclear power plants, at hydroelectric power plants, and in other technical systems (I. Aronov).

But in the transition to risk assessment in the *field of rare events*, the values of the hazard factors fall within the range of the probability density distribution functions (pdf) beyond the confidence intervals for reliability indicators—in the "fuzzy tail" area. Therefore, the applicability of some analytical formulas that are accepted as reliable and accurate, and models for estimating the probability values of events, falls into the area of pdf "fuzzy tails", which is unacceptable:

$$f_{f\,z\text{tails}} \Leftrightarrow \text{or} \rightarrow f(x), \tag{3.10}$$

where $f_{f\,z\text{tails}}$ are accepted as the Prdf with an undefined ("fuzzy") "tail"; *f(z)* is a clear pdf defined on the "hypercone of truth"; *or* is the operator of preference for the choice of the pdf function; $\Leftrightarrow$ denotes equivalence.

This is why ICAO documents, by default, recommend solutions to assess the significance of identifiable risks without using the concept of event probability, but indicating the need to use possible frequencies normalized in some way [6, 22] from the "*likelihood*" class [5] instead of the "*probability*" class.

It is this position of ICAO that substantiates the application of risk analysis matrices [5, 13] to perform expert estimations instead of the impossible accurate calculations needed when the term "probability" is used, which is virtually impossible in practice, due to the lack of accurate a priori data for pdf and Prdf.

An explanation of this situation can be given in the following way.

In the class of computational methods with Fuzzy Sets of initial parameters or characteristics of systems with pdfs from (3.8) that are fuzzy to some extent ($f_{f\,z\text{tails}}$), the problems break up into two classes:

- For dynamical random processes $x(t)$ ~ with continuous pdfs, the following correspondence is established for $x(t)$:

$$x(t) \sim f_k(t), \tag{3.11}$$

- For random processes with discrete pdf distributions of type $f_q(t)$ to change certain discrete states $q_i \in Q$, where the number of states $i = 1, \ldots, n$ is at countable, but almost always finite. In this case, each predicted process of changing discrete states is clear and is known in advance in terms of its properties, the number and type of elements in it and pdf. The randomness of these processes is only in the uncertainty of the time of occurrence of states, but in problems of investigating the event combinatorics this does not matter.

The pdf fuzziness $f_{f\text{ztails}}$ arises due to the fact that any calculated (e.g., analytic type f (z) above) pdf always display the results of experiments that contain processing errors and the uncertainty of information about the Prdf law due to the lack of statistical data. Therefore, the problem of the attributes of the "fuzziness" processes is solved in a special way, depending on the type of uncertainty and the way the pdf functions $f_x(t)$ are given.

3.4.2 Processes with Type 1 Pdfs ("Hard Tails" Type)

With clear pdf $f_x(t)$, under the assumption of their accurate description, only in a certain region (in the quantile—according to M. Fujita) exact (fixed) values $x(t)$ fall into the fuzzy region of small probabilities—the tail of the pdf distribution, where the pdf values (probability) can vary indefinitely in the range of 10^{-5}–10^{-18}—in the "heavy tail" region.

An explanation of this thesis is given in Fig. 3.1.

The best solutions were made by: G. Malinetskiy ("Heavy tails"), I. Aronov (indirect methods, confidence intervals), RAS–SST (fuzzy subsets, PSA update).

But the "*tail*" *area is not defined* due to the lack of statistics and "mistrust" even to the analytics. Any "*analytics*" is just a hypothesis accepted for truth, if the question of reliability is not considered in cases when it is not important in practice. So, it is

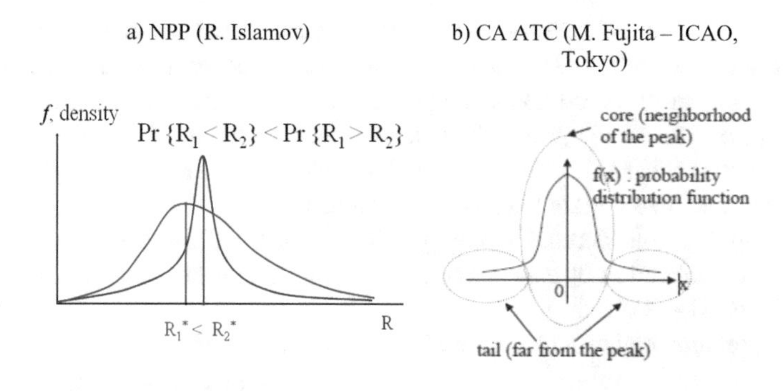

Fig. 3.1 Transition of "pdf" into "heavy tails"

accepted the actual value of pdf and Prdf in this area is unknown (usually attempts are made to approximate some histograms, unreliable, as a rule [15, 23]).

There are known attempts to refine the pdf approximation either on the basis of available statistics (Fujita [23]) or on the basis of "heavy tails" by Malinetskiy [11] in order, nevertheless, to calculate the probability of rare events. But *such calculations are no more than errors* and do not serve a useful purpose, since small values of 10^{-5}–10^{-18} cannot be distinguished in meaning and can only be *demonstrated as fictions. (Note by the authors.)*

In the risk theory, when assessing system safety levels (or flight safety), such an approach is unacceptable, which was also mentioned by Fujita [23].

3.4.3 Type 2 Pdf with "Fuzziness" of the Pdf Function

In this case, instead of one analytic function of the type 1 pdf from (3.11), two of its varieties are used, for example, in the form of pdf 1 and pdf 2 considered by R. Islamov as a result of a parallel shift of the pdf median along the axis of the argument x (Fig. 3.1).

Such a method is best known from the practice of calculating reliability indicators (in the RT) using the accumulated experimental data from aircraft failure statistics (in other technical fields, for example, at nuclear power plants, the same situation occurs).

In this case, the "*tails*" *method does not work.* It is necessary to take uniquely only one pdf, for example, pdf 1, which is sufficiently valid only for the Prdf values in the range up to 0.9999. Then, it is necessary to assume that the probability of a hazardous event in the risk region (FF, for example) should not be, more or less reliably, better than 10^{-5}–10^{-6} (it can actually be better, but this is *due to the fuzziness and cannot be proved*).

There is a paradox in the RT: The theory is rigorous, based on analytics, but analytics does not exist (does not work) in reality for the pdf regions in the "tails".

Thus, in order to estimate the probability of a rare event for other values of the process lying in the pdf "tail" area, this information is not enough. Any hypothetical and theoretical information about the pdf is undefined, and consequently, there is absolutely no sense in the probabilistic approaches. Such situations are recognized as the most correct in some areas of the nuclear industry (Arzamas-16, the Kurchatov BAS Center and, to a certain extent, in other areas).

For all options considered, the only justified way out is the transition to the Fuzzy Sets and the application of expert risk analysis matrices (per ICAO) (both per ISO and Gosstandart—R).

This is recommended and proposed in ICAO documents for civil aviation (in this book, this is the subject of research and an attempt to find a way out of the "conditional deadlock" in the RT).

Sampling Probability pdf with "heavy tail" by Monte Carlo

a1, a2 **– points for uncertain impulse of «heavy tail»**

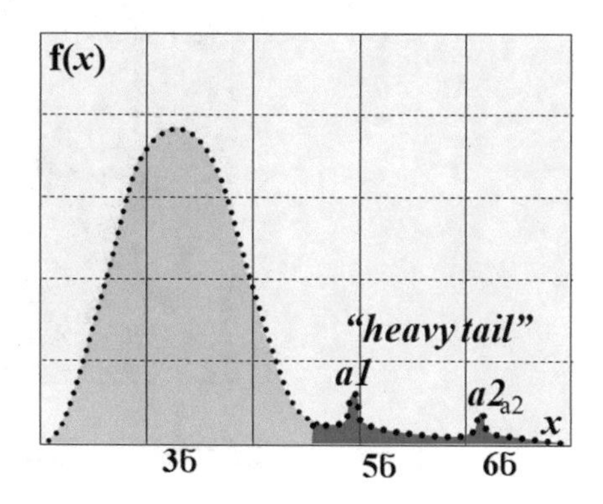

Fig. 3.2 Uncertainty in "pdf" for risk scenario

Thus, the *rare events problem is determined by the following factors*:

- Uncertainty of pdf and Prdf with fuzzy "tails";
- Uncertainty in the assignment of the boundaries of confidence intervals (according to I. Aronov).

The diagrams illustrating the manifestation of these factors were given earlier in Fig. 3.1, but received further support in NASA works (below), in particular in the form of pdf plot in Fig. 3.2.

3.4.4 Uncertainty of Pdf and Prdf in the NASA Experimental Results

Figures 3.2, 3.3, 3.4 and 3.5 show pdf and Prdf plots obtained in NASA [5] on the basis of simulation modeling with a large volume of tests. It can be seen from the figures that the "tails" are fuzzy, and the pdf values in the "tail" area are also fuzzy.

The results presented below are *demonstrated for the first time* to the broad scientific circles of the Russian Federation. (It can be assumed that experts in the probability theory could have known about this, but the authors have not succeeded in the search of the publications supporting this assumption.) *The results are borrowed correctly from [5] to prove here advantage of Fuzzy Sets.*

The value of these materials for confirming the RRS positions is quite obvious.

It can be seen that the values of the argument are "spread" so widely that it is impossible to find a clear Prdf quantile among the values in the "tail".

An extremely interesting (for the RRS) result is given in Fig. 3.3, where pdfs obtained by the Bayesian formula are compared.

Fig. 3.3 The priory and aposteriory distributions (altering of "pdf")

***d*– Initial pdf determined in PSA (calculated)**
***b*–pdf determined by Bayes (analytical - calculated)**

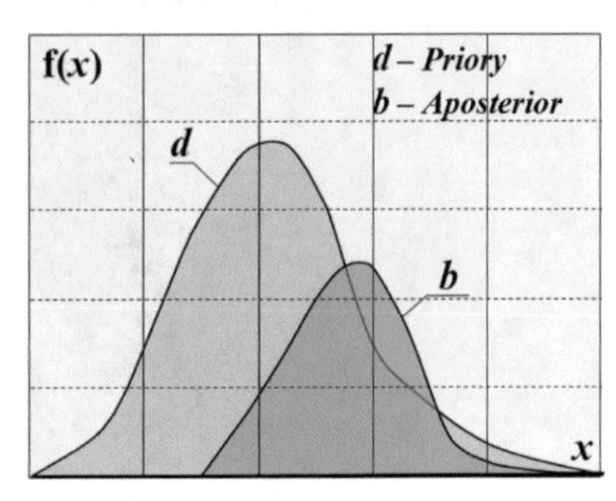

Fig. 3.4 The results according of NASA

Analytical Modelling

a) Analytical Smooth f(*x*) ∿ pdf according PSA

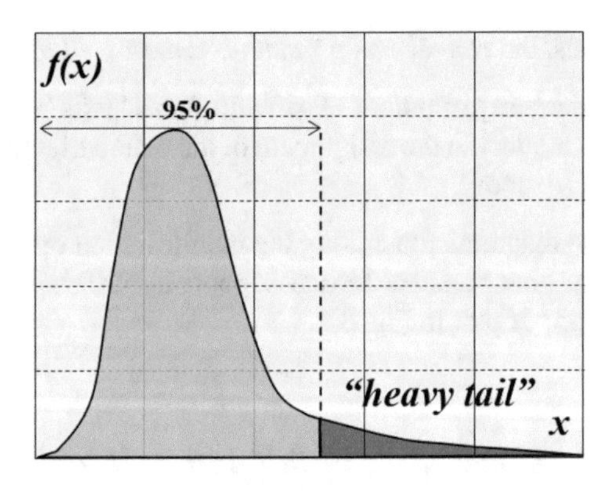

Fig. 3.5 pdf and CDF

b) Latin Hypercube by NASA for Input analytical Functions

Distribution Functions: *d* - Initial pdf (not true in PSA)
***b* - determined by CDF with quantiles of PSA**

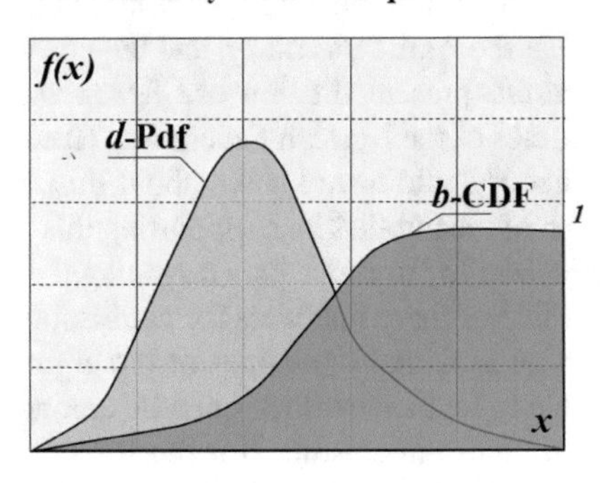

It can be seen that the pdf appearance has improved. But the “tail” has fuzzy forms, so the “improvement” by the Bayes method, which is promoted in CATS [8], is not achieved in the “rare events” problems.

The presented results confirm the special significance of the hypothesis about events with the “near-zero” probability adopted in the RRS. This can also be confirmed by permanent author’s comments in the CATS publication [8] on the non-informativeness of the pdf functions due to insufficient statistics (here, CATS means “cause–effect models” on Boolean truth lattices, as shown in [24]).

3.5 Transition to Fuzzy Sets from the “Boolean Lattice” in the RT

This section in the RRS doctrine is decisive for the development of methods for assessing the FS levels (or SS in ATSs in the broad sense of the word). Here, agreements have been made and evidence has been given of how to “reconcile” and combine the methodology of probability spaces and fuzzy estimates of risk levels in the rare events problem. The material is based on the conclusions from the previous Sect. 3.4 and the results of [1–3]. It suffices to indicate that on the numerical axis, any fuzzy probability value can be given a vector of fuzzy estimates. This is evident from the NASA pdf “tail” [5] in Fig. 3.2.

3.5.1 Initial Conditions

The problem is that systems in the RT are clearly defined on the Boolean lattice. This is applicable to all events in a structurally complex system. Consequently, any events must be measurable with the measure P from U (from 3.1). But the task is to assess the consequences for the ATS from the manifestation of hazard factors, which determine rare events such as functional failures. Taking into account that “rare events” (failures) are considered in a subclass of opposite events, the desired transition to the Fuzzy Sets can be performed in two phases.

One of the constructive proposals (phase 1) is a preliminary description of the predicted risk event R in the form of a random measurable event under certain conditions Σ_0 for defining a system in the probability space U from (3.1) in the usual sense [17, 18]:

$$R = A_*(\omega|\Sigma_0,U),$$
$$U = (\Omega,E,P|\Sigma_0), \tag{3.12}$$

where E is a “sigma-algebra”; P is a probabilistic measure (according to A. Kolmogorov [5]) of elementary random events $\omega \in \Omega$ from the outcome space Ω; A_*

is some critical event of the class type $A = f(E|\Omega)$. The criticality $(*)$ of events A made up of $\omega \in \Omega$ in $A(E)$ is determined by the consequences of functional failures, but without the probabilistic measure P_A. Probabilities P_{A*} "do not exist" (*or they do not make sense for events with the "near-zero probability"* with unreliable statistics).

Similar events A_* are found by the "event trees" by methods known in the RT [25, 26].

The key point is a FMEA-like program [25, 27], which is one of the main tools in the PSA method [26, 27]. In this case, the FS level is estimated through the criticality of the chains (from Chap. 1).

The allocated set of similar events $\{A_{x\,i}\}$ is strongly clear by physical signs, although the *measurability* by P from [17, 18] *cannot be attributed* to them due to their "rarity". *But their clarity is known*, because they are defined on a *Boolean lattice* (in the hypercube of truth per [2, 24]). It follows that a subset of the introduced events $\{A_*\}$ can be interpreted as a universal set of attributes for which predicates can be specified in the form of any supposed (in the predictive ICAO method) properties. It also means that it is *possible to introduce fuzziness of their significance* by randomness of appearance or by other signs, i.e., to create a fuzzy subset.

3.5.2 Solving the Problem of the SST Transition from the Boolean Lattice to the Fuzzy Sets

It is proposed to divide U from (3.2) into two parts [3] and introduce, per [1, 3], fuzzy subsets and "fuzzy algebra" $\underset{\sim}{\bar{A}}$ for events of the type $\bar{A}$ from (2.1):

$$U \rightarrow \underset{\sim}{\bar{A}}(\omega, M_R | \Sigma_0, \Delta E_R), \tag{3.13}$$

$$\Delta E_R = E \backslash \Delta E_*, \tag{3.14}$$

where ΔE_R is the area of events (subset) clear in function; $\Delta E_* \subset E \subset U$ is a set of critical events (functional failures); M_R is a set of measures for fuzziness, randomness, and significance of critical functional failures taken from the universal set $\bar{A}_*$, taking into account fuzzy estimates like: "low risk", "high risk", "significant risk", "acceptable risk".

A *fuzzy measure* of the random occurrence of a risk event R will be as follows: "rarely", "very rarely", "frequently", "infrequently", "sometimes", which allows the use of known ICAO risk assessment matrices from the FAA, MES, etc. This makes it possible to avoid the incorrect concept of *"predictable" ("guessed") probability of an event*, as noted in [7, 14, 28]. To illustrate this, the relations from [14] are given below for the mathematical representation of the ICAO idea in the form of (2.13) (from Chap. 2), but with the fuzziness index. Thus, these formulas are identical in form to those given in Chap. 2, but they are completely a different fuzzy, since they introduce the values of an integrated risk assessment:

$$\underset{\sim}{\hat{R}} = \left(\underset{\sim}{\mu_1}, \tilde{H}_R|\Sigma_0\right), \quad \underset{\sim}{\hat{R}} = \hat{f}\left(\tilde{R}|\Sigma_0\right) = \tilde{f}\left(\underset{\sim}{\mu_1}, \tilde{H}_R|\Sigma_0\right), \tag{3.15}$$

where $\underset{\sim}{\hat{R}}$ is an integral assessment of the risk level (the "amount of hazard" as suggested above) as a function of the two-element set, but a fuzzy one.

In this case, the concept of a *scalar—"average risk"* with rare events—is now correctly (automatically) completely *excluded* from consideration. This is difficult to do in the PSA, which was proved in Chap. 2.

Indeed, in problems of this kind, *calculating the scalar as a measure of risk in the class of events with the "near-zero" probability leads to a complete absurdity* in the interpretation of the physical meaning of the questions under study. (These comments are given in the book [18] by authors V. Korolev, V. Beningen, etc.)

(These results were reported at one of the STC meetings with the participation of RAS Associate Member N. Makhutov).

3.6 General Scheme for Constructing Fuzzy Risk Models in ATSs

Possible alternative directions and principles of constructing models for analyzing the quality of functioning are either the RT or the SST [4, 6, 29]:

$$X \rightarrow M_x \overset{(D_x, B_x)}{\Rightarrow} \mathrm{RT}(\mathrm{PSA} \Rightarrow \mathrm{SS}) \tag{3.16}$$

$$X \rightarrow \underset{\sim}{X} \overset{\left(D_x, \underline{B}_x\right)}{\Rightarrow} \underset{\sim}{\left(A_x, \underset{\sim}{M_x}\right)} \Rightarrow \mathrm{SST}(\mathrm{SS}, \mathrm{RT}) \tag{3.17}$$

where the lower index (...) denotes that here fuzzy subsets $\underset{\sim}{X}$ of a clear universal set X (or another kind) are considered; M_X denotes characteristics of subsets formed on the basis of the "hypercube of truth"; $\underset{\sim}{X}$ is a fuzzy subset created on the basis of the initial universal set X some assigned fuzzy properties; $\underset{\sim}{A_X}$, $\underset{\sim}{M_X}$ are various fuzzy subsets determined by the conditions of the problem and the properties of the technical systems for which "*hazard models*" are developed, for example, by ICAO, on the basis of the methodology for calculating risks.

Thus, the same universal *clear set* X generates two types of system models in fuzzy subsets $\underset{\sim}{X}$: clear (in the RT) and fuzzy (in the SST) ones. Below are given notations for some transformations:

B_X is the Boolean basis for clear subsets M_x formed from a clear universal set by applying logical operations to establish the "truth" of the statements obtained.

$\underset{\sim}{D}$ is the combinatorial basis of the system constructed taking into account the properties of specific connections of reliability elements.
$\underset{\sim}{D_X}$ is, accordingly, the designation of procedures for determining the combinatorics for connections of reliability elements of the functional structural diagram of the system, but on the basis of fuzzy subsets (3.17).

3.7 Analysis of the Basic RT Provisions Determined by the Hypothesis of the Existence of a "Hypercube" of Truth for Objects from Clear Sets

In the RT, as shown in Chaps. 1 and 2, the system operability in a given sense [14, 16, 26] is determined by the structural diagram of connections of the physical reliability elements $e_i \in E$ on the example of Fig. 3.6 from Sect. 3.9, which describes the features of this approach. Here, the results of the corollary analysis are given that follow from the hypothesis of the "hypercube of truth" in the RT when constructing risk models in the sense accepted here (without "probabilities").

The properties of such a system with the achieved standard reliability indicators [16, 24, 26] are usually estimated with the help of some performance function (PF). The value of this PF function is measured either through the probability of system states or through the values Y of the logical algebra function (LAF) when logical analysis procedures are used for discrete states of the system [8, 24]. For the PF form estimated by the probability of occurrence of the corresponding states, as is known [26] indirect performance indicators of the system are also specified in the form of the standard performance probability or time t_{ot} operating time to the first failure, availability factors, repairability, etc. (for non-recoverable systems, in particular).

It is assumed that the set E of physical reliability elements is given (finite or countable). The first assumption is that a finite (limited, countable) set $E = \{e_i | i = \overline{1, m}\}$ of elements $e_i \in E$ numbered in some order $i = \overline{1, m}$ with a predetermined known number $m \ll \infty$ is defined.

At the same time, a corresponding structural diagram for connections of reliability elements is developed and established, and this combination determines the physical basis or system resource necessary to ensure the functioning of the specified system:

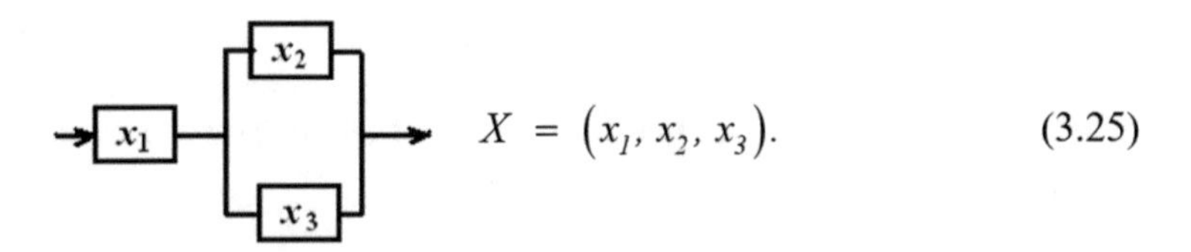

$$X = (x_1, x_2, x_3). \tag{3.25}$$

Fig. 3.6 Example of a parallel connection of reliability elements

$$\{e_i\} = E \equiv \left\{e_i | i = \overline{1,m}\right\}. \tag{3.18}$$

This basis is the primary prototype *(the original) of the future system*, where the resource provides the primary, most important functional physical properties of the system as follows:

"Signal transmission in the system" (3.19)

The property (2.20) ensures the functioning of the system as a whole (in the class of systems defined with both clear and fuzzy subsets). This property as primary can be specified by some special functions of the quality of "functionality" [10, 30, 31].

Accordingly, another opposite property is possible in the system:

"Functional failure". (3.20)

The property (3.20) is determined by a set of other possible functional states of the physical elements $e_i \in E$, when the system "*does not transmit the signal*" (with z_i—for elements with a logical negation of the "signal transmission" property of the element x_i):

$$A \rightarrow \bar{A}. \tag{3.21}$$

The main initial RT hypothesis with respect to the physical properties of E that determine the quality of the system in the form of (3.19) or (3.20) is accepted as the *global RT postulate* on the independence of the considered elements as:

"All $e_i \in$ E work independently". (3.22)

This means, for example, that according to (3.22), the failure of the aircraft actuator due to a defect in the feedback transmitter does not affect the gearbox failure (on the same aircraft) due to fatigue failure of the gear cog, etc. Further, the stabilizer control button on the IL-86 can be "mistakenly" blocked by the PIC, which leads to an accident ("M. Neymark's button" [27]).

$$T : V \rightarrow [0,1] \Rightarrow T(P \supset Q) \rightarrow [0,1)$$
$$T(P \supset Q) = \max\,\{min\{T(P),T(Q)\}, 1 - T(P)\}$$

3.8 Basic Provisions of System Models in Fuzzy Sets

The physical prerequisites noted above allowed for introduction of two basic provisions in the classical RT as follows:

Provision 1. A clear set of elements of a system type E (LAF) or $A \supset X$ defines an abstract mathematical model (original) in the form of a *clear universal set* for describing the physical basis of the system. The system consists of a set of elements E_s defined in the form of a set X of special elements x_i that take only two logical values "0" or "1" and denote the discrete states of these elements:

$$x_i \in X; x_i = (0 \text{ or } 1), \{x_i\} = X, \tag{3.23}$$

where all x_i are independent in the sense of (3.23), but belong to the set of elements (3.23) taken in (2.24):

$$\{x_i \sim e_i\} \Rightarrow X \sim E. \tag{3.24}$$

It means that the use of the concept of a "*Universal (clear) set*", for example, in the form of X on the basis of the hypothesis "on the hypercube of truth", led to the creation of a modern classical RT with all its branches, including the PSA, and caused incorrectness in solving the issues of the "rare events" problem.

Provision 2. In the universal clear set X, all elements $x_i \in X$ are independent in the same way as in the physical basis E_S. Each element can alternatively have one of two values: "0" or "1". This universal (clear) set (3.18) can be used to construct some models (or theories) of the system functioning. For example, this is suitable for those ATSs where assessments of the system operability are found with some criterion—the system safety (not reliability).

3.9 Algebra of the Events Logic in Catastrophic Scenarios

3.9.1 General Provisions Determining the Nature of Catastrophes

In the general RT and SST relationship, it was shown that in the RT, it is not possible to properly estimate the SS level in special cases, since the RT, due to the reduced probability of functional failure (FF) occurrence, provides only the recommendation for the catastrophe occurrence time to be "*moved to infinity*" as events of the functional failure type from the set of states defined in the sets of the minimal cut sets of failures (MCF) [16, 26]. For example, according to the SST, the example "Helicopter with a single rotor" mentioned in Chap. 2 is "hazardous", since a "catastrophe" is designed in it: If a breakdown of the main rotor (or swashplate or gearbox) occurs, i.e., the risk of negative consequences is high. The probability of such an event is negligible (and even unknown in advance), but such a negative scenario is clearly predicted.

Aircraft IL-62 (the most reliable civil aircraft in the "past times") with four engines is, following the SST positions "very hazardous" during take-offs, since if two (and even one) engines fail at altitudes up to 1000 m, the remaining two—three engines do not ensure a normal emergency landing from such an altitude.

3.9.2 Use of Logical Algebra Functions (LAFs) for Evaluating the System Operability in the Reliability Theory (RT) and in the SST for the Construction of J. Reason Chains

The use of LAFs [24, 32] is convenient for defining scenarios for the development of hazardous situations, which is necessary in evaluating the conditions for the system health based on the performance functions (PFs) in a logical form. This makes it possible to further find the system operation indicators in fuzzy subsets (3.15).

With known LAFs, discrete events and combinations of element status indices can be determined that characterize the system operation, its state and standard indicators of system reliability. The proposed method is more universal, since it can provide, if necessary, a transition to the traditional PSA, if areas of certainty (reliability) of the results obtained are known. The main purpose of the LAF methods is to search for the conditions for the occurrence of catastrophes in the form of some scenarios and to construct J. Reason chains leading to a catastrophe [13, 33].

In the scheme proposed below, the transition to fuzzy logic of events makes it possible to relate "*logical equations of the conditions for occurrence of catastrophes*" to the structure of J. Reason chains. This is necessary when constructing, for example, ICAO hazard models (for J. Reason chains). Here, it is presented in a formalized form for the first time, whereas in the ICAO documents, it is just a heuristic scheme in the form of a "cheese chart" [34, 35] resulting from common sense.

On the basis of the "hypercube of truth" hypothesis, the universal set implies (3.23) or (3.24) two ways of constructing clear subsets as two tables in the form of codes and in the form of LAFs for given connections of the reliability elements. In this case, the LAF value equal to unit will denote the system operability on the basis of "signal transmission" criteria in a specific scheme for connecting the reliability elements (i.e., the PF—the system performance function).

It is important to note that the proposed approach was to some extent envisaged in the RT in the form of an event tree method, but it was rigidly oriented to calculating probabilities using the hypothesis of the "hypercube of truth". Therefore, there were difficulties in assessing the event safety levels in connection with the "rare events" problem. Only the development of this method in the form of J. Reason chains or "Canada-based" scenarios made it possible to circumvent theoretical difficulties in the RT [26, 36, 37].

As an example, a simple series–parallel connection of elements is shown (Fig. 3.6). A universal (clear, finite) set X for the parallel connection of reliability elements in Fig. 3.6 is as follows:

Table 3.1 (**a**) Codes ~ d_i; (**b**) LAFs ~ yx

(a)					(b)					
	x_1	x_2	x_3	q_i	X_k	x_1	x_2	x_3	$y\,(X_i)$	
d_1	0	0	0	q_0	X_1	0	0	0	0	Norm
d_2	0	1	0	q_1	X_2	0	0	1	0	Norm
d_3	0	1	0	q_2	X_3	0	1	0	0	Norm
d_4	1	1	0	q_3	X_4	0	1	1	0	Failure
…	…	…	…	…	X_5	1	0	0	1	Failure
				…	X_6	1	0	1	1	Failure
…	…	…	…	…	X_7	1	1	0	1	Failure
d_8	1	1	1	q_7	X_8	1	1	1	1	Failure

The formalization of the procedures (3.23), (3.24) to find clear subsets $X_i \subset X$ for the universal set X in a particular example is given sequentially in accordance with the scheme (3.26) initially in the form of a set of two tables: Table 3.1a, b: first (a), then (b) [4, 38].

The first variant of constructing (clear) subsets X_i composed of a recursive alternation of the values "0" and "1" makes it possible to determine the d_i codes in each row of Table 3.1a. Formally, the sets d_i and X_i consist of the same numbers. But in Table 3.1a, the d_i symbol is the designation of the code (3-digit), and X_i—in Table 3.1b—is a set of indices: "*true-0*", "*false-1*".

Elements $x_i \in X$ that take logical values of "1" in the failure state can be re-denoted to new elements z_i through logical negation operations, i.e.,

$$z = \bar{x}_i\,(\text{i.e., “failure”}). \tag{3.26}$$

The result of mapping the set of combinations of discrete values "0 or 1" for each of $x_i \in X$ in (3.25) with dimensionality $Dim[X] = m$. m is the number of physical elements $e_i \in E$ from (3.24) that gives a set m of bit codes as Table 3.1a. This determines a certain number of combinations of state signs x_i in the universal set X that specify the combination code, as well as the discrete state $q_i \in Q$ with the given code d_i. The number of states $n = 2^m = 2^3 = 8$, the number of codes $d_{i,}$ is the same and corresponds to the number of discrete states $q_i \in Q$.

The combinations considered are given in Table 3.1a. This result reflects the RT methods of investigating systems with discrete states by means of various models of the functioning processes [11, 15, 26, 39].

Since it was assumed in Table 3.1a, as an example, that X is a set of three elements $(x_1, x_2, x_3) = X$, the actual numerical value of the code can be recognized as a symbol of a given state of the system $q_i \in Q$, as is customary in the RT [14, 26]. This is the benefit of this form of presentation of the clear subset X_i already in logical form and in another Table 3.1b for the given universal set X.

Further, on the basis of the same Table 3.1a, a complete Boolean address is formed by adding a set of logical algebra functions (LAFs) to the Table 3.1 for a particular connection of reliability elements adopted here, for example, for Fig. 3.6 (3 elements) in the form of a logical function:

$$y_x(X_{m\,\Sigma}) = f_L(X_{m\,\Sigma}) \equiv y_{x\,\Sigma}(\{X_i\}), \tag{3.27}$$

The values in (3.27) are as follows:

$$y_{x\,\Sigma}(\{X_j\}) = f_{L\,\Sigma}(\{X_j\}) \tag{3.28}$$

In this case, $y_{x\,\Sigma}$ is a LAF in the entire set $X_{m\,\Sigma}$ composed of subsets $X_j \subset X_{m\,\Sigma}$ from Table 3.1b. Here, X_m is a set of values of "0" and "1" in the rows of Table 3.1b. But these indexes, which determine the bit value in Table 3.1a, are now interpreted in Table 3.1b as logical values of "0" and "1".

In the example considered (for $n = 3$), for example, if a connection $e_i \in E$, $i =$ 1,2,3 is given, then the system is operable in the last three states, where the "ones" stand.

Therefore, $Y_x(x) = f_L(x) = 1$, i.e., the LAF, according to the rules of logical algebra, will have a value of one, "*truth*", which in the PF form means "signal transmission", etc. Such an operation makes it possible to eventually obtain a set of "logical words" represented in Table 3.1b as $X_{i.}$ This yields numerical values of the logical variable as the values of the indices: 0 means "norm" (signal passing), 1 means "failure" (these symbols are introduced for "inversions", which is more suitable for estimating the FS level through "hazards"–"risks" as in [6]).

The desired logical disjunction of conjunctions will be as follows:

$$y\ (X_3) = x_1^1\, x_2^0\, x_3^1\ \vee\ x_1^1\, x_2^1\, x_3^0\ \vee x_1^1\, x_2^1\, x_3^1 = x_1\, \bar{x}_2\, x_3\ \vee x_1\, x_2\, \bar{x}_3\ \ \vee x_1\, x_2\, x_3, \tag{3.29}$$

where X_i is a subset of elements of the logical word on which the logical value for the LAF is determined as a conjunction. The logical variable in (3.28) is taken in the powers of "0" and "1" according to [18]. This step allows for algorithmic description of the "negation" operation.

If in the example considered a certain connection of the elements is specified (as in Fig. 3.6) $e_i \in E$, $i = 1,2,3$, then the system is operable in the last three states, where the "ones" stand. Therefore, $Y_x(x) = f_L(x) = 1$, i.e., the LAF, according to the rules of the logical algebra, will have a value of one, "truth", which physically denotes the PF as "*signal transmission*", etc. The main property of the PF from Table 3.1b is its uniqueness. Therefore, the PF value on the hypercube of truth is taken as a truth criterion, for example, when checking the equality of two LAFs [24].

The integral LAF includes the total LAF number (for Table 3.1b) and will be a disjunction of all conjunctions in the rows of the table where FALs are shown:

$$y(X_m) = \bigvee_{j=1}^{n} K_j = \bigvee_{j=1}^{n} K_j \bigwedge_{i=1}^{m-r} (\bar{x}_i \vee \bar{x}_i),\ \bar{x}_i \in K_j. \tag{3.30}$$

The conditional concept of "incorrectness of the methodology for assessing system safety levels" on the basis of the PSA approach for rare events, on the one hand, is a consequence of the rigid application of the approach to the estimation of standard RT indicators in the form of LAFs in (3.30), (3.31). On the other hand, this is the result and consequence of the adopted hypothesis that there exists a universal set X in the form (3.23) for the physical basis E in (3.24).

Comment: The truth of the statements in terms of the hypercube (in the form of a Boolean lattice) "causes", or rather allows (which is more important in practice) *multiplication* of the "*probabilities*" of independent events, as is customary in the RT on the basis of the theorem in [24].

However, such a thing is possible, if all the pdf, Prdf are defined (accurately, analytically). This implies the need for a requirement to ensure the actual existence of this condition within the hypothesis of the "hypercube of truth".

This is a wonderful property, but it leads to a "*deadlock*" in the RT, since almost all known pdf and Prdf only approximately reflect real properties in some confidence limits. And the "tails" of such *functions are indefinite and "fuzzy"*. Therefore, the probability values of rare events with the order of 10^{-6}–10^{-20} should be considered inaccurate and cannot be trusted. It also follows from this that the use of the PSA leads unconditionally to *incorrect conclusions, if on the basis of the PSA "it is possible to formally obtain" small values of the probability of a rare event by multiplying the fuzzy "false" analytic pdf* in the range of ~10^{-6}, 10^{-8}, etc.). Formality lies in the fact that the pdf "tails" are *erroneously associated with the meaning of "truth"* adopted in the analytic regularity, which in fact does not exist because of the lack of reliable data.

The removal of the contradictions in the theoretical positions of the classical RT manifested in the PSA method in the SST transitions to rare events is ensured, if one recognizes the need to analyze the properties of rare events on the basis of the same concept, the so-called universal set $X \sim E$ with the properties of (3.24).

The fact is that these *properties of the "rarity of events" are not expressed in any way through LAFs* and *do not even characterize* the system *operability through "probabilities of states"*, since the probability of rare events is "near-zero". Elimination of contradictions is a transition to the new RRS doctrine formulated here in Chap. 2. Abandoning the concept of probabilities of the occurrence of random events falling into the region of pdf "tails" is unavoidable, as the PSA has nothing constructive virtually to *eliminate the unreliability of the results* with a low probability of the element failure. At the same time, as it was previously emphasized, that the RT is a quality base of technical systems, without which, obviously, it is useless to create any systems at all.

That is why in the paper cited here [26] two functional modules were proposed for the methodology of system quality assurance by technical properties ("*structural safety*" by factor F1 in the RT) and "*operational safety*" (by factor F2 in the SST).

The provisions ("Boolean lattice", etc.) in the RRS doctrine make it possible to harmonize the RT and SST methods in the study of phenomena with rare events. The basis of this position of the "*hypercube of truth*" is the replacement of the positions of the Boolean lattice [24] by the "*hyperspace of vectors*" [2] (the "*vector lattice*") defined in the fuzzy subsets $Q_m \subset \underset{\sim}{Q} \sim X$ given in (3.24) above. Taking into account the comments made, the formalization of the RT provisions can be given in the form of descriptions of the classical (clear) RT in the class of systems with some fuzzy properties of objects from fuzzy subsets of universal sets from Sect. 3.9.

3.10 Positions of the Classical Reliability Theory Based on the Hypercube of Truth

3.10.1 Universal Method for Formulation of the Classical Reliability Theory Fundamentals Using the Fuzzy Sets Positions

The RT fundamentals are given, for example, by Aronov [26] and others [40, 41]. Here, the original difference in the RT formulation lies in the fact that additional concepts of some universal set of elements E [35] have been preliminarily introduced. Further, in accordance with the RT rules, as in [30, 42], the consequences of this provision are studied. This corresponds to the construction of system reliability models in "clear subsets" $A_i \subset E$.

The main recommendation is a reference to the need to use the notion of "hypercube of truth" of the corresponding statements. This approach implies *strict restrictions on the properties of the elements* $e_i \in E$, in particular on the *independence of events such as physical failures* of elements. Then many models accepted in the RT classification over the last 50 years of its development become clear. This led to the creation of famous schools [16, 24, 26, 41] in this field and ensured significant success in the theory and practice of creating highly reliable systems. This is the achievement of the classical RT.

However, the traditional approach discussed in this subsection leads to some difficulties in the development of the system safety theory (SST), where the key point is the study of the properties of rare events such as an accident.

The SST approach makes it possible to obtain more correct results than in the PSA method, by applying the *Fuzzy Sets* methods of [1, 4] to study the properties of rare events.

3.10.2 Initial Hypotheses of the Classical RT Defined on the Hypercube of Truth (on Boolean Lattice)

It is known [43, 44] that clear provisions and methodology of the classical RT can be equivalently stated from positions of fuzzy subsets [1, 2]. Then, the transition to the system safety theory (SST) for complex systems on the basis of risk concepts in the field of "rare events" is fairly simple and logical.

It is assumed that in (3.18), physical reliability elements $e_1, e_2, \ldots$, etc., are set, with which a structurally complex (according to I. Ryabinin) functioning system is constructed. The indicated sets of elements e form a universal set E of physical (clear) elements:

$$E = \{e_1, e_2, \ldots, e_n\}.$$

Based on (3.24) introduced above, a set X of logical reliability elements $x_i \in X$ is formed:

$$x_i \in X, x_i \sim e_i \in E \Rightarrow X. \tag{3.31}$$

Logical elements are just designations of possible physical elements e_i such as "failure", "non-failure" that characterize physical states and mathematical (logic) states of elements as follows:

$$x_i = (0 \text{or} 1), \text{“0”means“norm”}, 1\text{“means”“failure”}. \tag{3.32}$$

By the statement of the problem, all the elements of the universal set E in (3.22) are clearly independent, since in essence it is a "*set of cubes in a box*". It is this provision that is the basis of almost all models in the RT.

Thus, it follows that all the elements x_i or e_i of the universal sets X and E can be assigned clear properties in the form of probabilities p and q—failure or non-failure in any considered time interval or in a given situation in the form of some conditions:

$$x_i \in X_\exists(p_i, q_i); p_i + q_i = 1. \tag{3.33}$$

This condition (3.33) is fundamental in the RT and, on the one hand, allows constructing rigorous probabilistic models of systems, but with restrictions on E from (3.32).

The analysis of the functioning of systems is based on the "*idea of the hypercube of truth*", etc., but on the other hand, leads to the incorrectness of individual PSA results in situations with rare events, which was shown above. For example, on the "hypercube of truth" [24] models of "monotonicity" of the probability functions are defined by I. Ryabinin and I. Aronov provided that the Borel–Cantelli–Kolmogorov theorem and others are met [17, 18]. On this basis, in a given "failure tree" or "event tree" (according to NASA [5], B. Gnedenko) and others (according to A. Kashtanov or

E. Barzilovich [19, 41]), probabilities of events are calculated within the *hypotheses of the "hypercube"*.

This is a great advantage of the RT methods, since within the framework of the formulated provisions, structural connections of elements can be constructed that provide only "standard reliability". This makes it possible to find graphs of transitions in the "birth-and-death process" and to study properties of recoverable and non-recoverable systems for sets of technical facilities with various actual operating times during the product life cycle [18, 41].

This is *important for ensuring the quality of systems*, but it is *not possible to solve properly the problems of ensuring system safety* with this approach. In this connection, the RRS doctrine was proposed in Chap. 2.

3.11 Determination of Paths to a Catastrophe Using the "Hypercube of Truth" Model for Values of the State of Physical Elements of the System from the Universal Set

3.11.1 Nature of the RT Postulates on the Independence of the Change in the State of Physical Elements of the System

The basis is the hypothesis of the difference in the physical properties of independent reliability elements of the system connected in some structurally complex schemes, as shown in Fig. 3.6.

Logical states of these elements such as "non-failure" or "failure—loss of functioning properties" are coded with the following values: "0—*norm (non-failure)*", "1—*failure*" (in inverse form via logical elements $z_i = x_i$). The same characteristics adopted in the RT are introduced in the Fuzzy Sets theory.

The RT assumption on the existence of a hypercube of truth properties with Karnaugh maps on the basis of the Venn diagram [24] makes it possible to construct a strict axiomatics of the RT positions. However, this RT axiomatics resulting from the properties of the hypercube leads, as was shown above, to a "deadlock" in problems with rare events [5, 26, 45].

The orderly positions of the classical RT (I. Aronov, I. Ryabinin [24, 26]) allow for application of the following general postulates of great practical importance:

- assignment of a pair of numbers (p, q) (in Fig. 3.2) in the form of probabilities (failure, q or non-failure, p), for each independent physical element from a given set E, $e_i \subset E$;
- construction (as in Fig. 3.2) of any (conditional arbitrary) structurally complex connections of the reliability elements from the same elements $e_i \subset E$.

Consequences of the formulated provisions:

(a) If the "hypercube of truth" is used in the analysis of structurally complex systems, then it is possible to search for the true "*minimal cut sets of failures*", which makes it possible to determine the probability of functional events using the PSA-based "tree" method (but it is impossible practically and strictly theoretically in "rare events problems").
(b) If the domain of "*fuzzy subsets*" of some universal set of attributes is used, then the "*minimal cut sets of failures*" (resulting from the hypercube properties) can be used to construct J. Reason chains without using probabilistic indicators of events, but using *fuzzy risk indicators*, as was demonstrated above.

In this case, the combinatorics of rare events can be analyzed by computer methods without using a probabilistic measure of events, which is quite obvious if the system is highly reliable.

Below are examples of the application of the *ADS* program (*DEMO* version for analyzing combinatorics of events. The results of the construction of J. Reason chains in the class of *MCF* sets and the corresponding state change graphs are given in figures of Chap. 5 below).

3.11.2 Logical Equation of a "Catastrophe" (According to I. Ryabinin) for Events from Clear or Fuzzy Subsets

The logical formula of the conditions for the occurrence of a physical event—a "catastrophe"—is based on the formula that a functional failure of a catastrophe type is a critical "*disjunction of conjunctions*" of the logical signs of selected events only, including also failures (factors and reactions) that together can lead the system to a *critical state in the minimal cut set of failures*.

According to Ryabinin [24], the probability $P(Q_*)$ of a hazardous class of discrete states $Q_* \subset Q$ of a system equivalent to the occurrence of a "catastrophe" event is formally defined simply as the probability P of a class or subsets of events only Q_* from a probability space U.

It is known [17] that the "*probability*" operator can be applied *only to events* in certain probability spaces, so the following can be written:

$$P(Q_* \subset E \subset U) \Leftrightarrow P(Q_* \subset E \subset U | Y_S(Q_*) = 1) = P_{Q*}, \quad (3.34)$$

where $Y_S = 1$ is some (logical) "statement". Here, the *FNRS* interpretation by I. Ryabinin is given, for which in the RT one need to find probabilities on the basis of the "hypercube of truth" properties.

But in the STT, the *concept of probability for a "catastrophe" is not required*, as a "catastrophe" as an event is estimated as the probability through the upper limit of values in the range of ~10^{-6}. Therefore, the theory of risks and damage assessment according to the ICAO methodology (Annex 19) is applicable. Thus, due to the combinatorial analysis of events, only the logical equation of a catastrophe can be found in order to associate the damage assessment procedure with the functional failure.

The main error in the methods of assessing the system safety based on traditional RT approaches for solving rare events problems is the incorrect construction of a discrete outcome space in the testing of complex systems. In many cases, discrete spaces, especially in civil aviation, are defined using a set of elements that are not a parent entity. This usually leads to violation of the conditions of the Borel–Cantelli lemma [17, 18]. Therefore, the key point in the SST is *to search for a set of event chains in the form of a critical "disjunction of conjunctions" from the MCF, and then to transit to J. Reason chains* [10, 13, 31] without probability indicators. This approach makes it possible to find the answer to how to solve the problem of reducing the predicted risk of an adverse event to an acceptable level.

3.11.3 Concept of Constructing J. Reason Chains in Fuzzy Subsets of States in the SST Using the FMEA and CATS Approaches

The concept of fuzzy subsets provides harmonization of various approaches to solving rare events problems (in both the RT and the SST).

The contradictory nature of the RT and SST positions in assessing the significance of catastrophe risks consists in the fact that the initial conditions for the occurrence of catastrophes are determined in the RT with the "logical equation of a catastrophe" (for MCFs) on the basis of constructing the logical algebra functions (LAFs), but only for the hypercube of the truth for the system element states.

But in the SST, *J. Reason chains and the risks of chain significance are also determined for the LAFs found in the RT, but in the class of fuzzy subsets in a vector hyperspace and without the use of probability indicators*, as highlighted above.

According to ICAO SMM [21] and on the basis of the Annex 19 recommendations taking into account by the SARPs, it can be established that J. Reason chains are nothing more than the result of applying the standard FMEA program at the structural level. Ilyushin Design Bureau ("IL Concern", experts M. Neymark, etc.) also presented a FMEA program—the baseline for ensuring the reliability and safety of aircraft based on the FF analysis. In this case, the FF criticality tables should be applied. *But such tables do not provide a solution to the rare events problem.*

Typical RT LAFs defined on the hypercube of truth have a typical form that can always be converted into a non-standard SST form based on the algebra of the fuzzy subsets logic. Recommendations are as follows:

$Y_S = 1$—the *"disjunction of conjunctions" of the MCF state sign* of the type "failure–non-failure" specifies *"failures as events of functionality property loss"* in a given structurally complex system in the RT.

Thus, in the LVR method [46], only on the basis of the "hypercube" of truth properties *certain LAFs for the PF are taken as unconditional events*, which can lead to incorrect results, since it is more properly to determine hazardous events in the σ-algebra of events from [36, 47].

However, the structure of the probability space in the *LVR* method [24] is not considered and is not reflected in any way, *since it is practically impossible with the help of hypotheses about the "truth of the statement" from the hypercube*. Certain results can still be obtained with the *use of concepts of conditional events and conditional probabilities*, but only for the case of simple structural connections of reliability elements.

In contrast, according to ICAO, J. Reason chains are considered as the basis for calculation and logical operations and procedures for identifying the causes of catastrophes. It is advisable to adopt the functional SST module from Sect. 3.1 of this book (paragraph 1.2.3) as the basis of the methodology for calculating risks, for example, in civil aviation, for nuclear power plants or hydroelectric power plants, and to develop procedures for fuzzy estimates.

The use of the concept of clear measurability of random events is only necessary at the first step in the development of highly reliable systems in order to exclude a special class of catastrophes with "game uncertainty" from consideration and not to solve the "rare events" problems, according to ICAO.

However, for events with the "near-zero" probability, it is necessary (and possible) to ensure the transition within the SST (not the RT) to fuzzy measures of risks and safety levels in relation to selected events falling into pdf tails (*according to* Fujita [23]).

3.11.4 CATS Concept (ICAO—"Netherlands")

The CATS general concept is just the FMEA, but it has some procedures developed for performing a formalized logical analysis of reliability schemes using Boolean variables. This is an achievement, but the same incorrect recommendations are present here as in the NASA plots: The CATS should use pdfs with good "tails" that do not exist (and this is directly stated in CATS [8]).

3.12 Formalized Models for Assessing Reliability and Safety of Systems with Discrete States

Models of systems with discrete states reflect real ATS processes more properly. Indeed, any failure is a discrete (stepwise) change in the state of a system.

Exactly, these terms describe here the well-known state change analysis programs that lead to the "event tree" in the FMEA and J. Reasons chains, and that ICAO recommends for the creation of flight safety or aviation safety management systems. In this chapter, the models discussed are necessary to substantiate methods for assessing risks as an amount of hazard measured in new-generation SMSs without probabilistic indicators.

3.12.1 Initial Definitions of the S System

The model of an S system with discrete states $q_j \in Q$ has the following form:

$$S = \{X, Q|G, \xi \in \Omega_\xi, T, \Sigma_0(\Phi)\}, G = (G, \Gamma Q) \tag{3.35}$$

In (3.35), a set $X = \{x_i|i = \overline{1,N_x}\}$ is introduced as a set of elements connected in accordance with the connection structure for the reliability elements, as is done in Sect. 3.7. Here, $X_\phi = \{\phi_k|k = \overline{1,N_\phi}\}$ are hazard factors affecting the state of the elements $x_i \in X_\varphi$ with indicator χ_i on the Boolean lattice:

α_1: $\chi_i = (0 \vee 1) := 0$—“norm” (“non-failure”);

α_2: $\chi_i = (\chi_i = 1) \equiv x_i$—not “norm”; $x_i = (\chi_i = 1) = \bar{x}_i$;
α_3: $\overline{x_i} \equiv z_i$—“failure”, $\overline{z_i} \equiv x_i$;
$z_i \in Z \equiv \{z_i \mid i = \overline{1,N_z}, N_z = N_x\}, Z \Leftrightarrow X$.

Next positions are demonstrated here:

$Q = \{q_j \mid j = 1,2,\ldots, \Omega_{\xi j}$ is the space of discrete states; ΓQ is a mapping of $Q \to Q$, i.e., a set of arcs or edges joining vertices in graph G.
q_i is a discrete state: $q_i \in Q$, defined as a combination of N_x elements x_i.
$x_i \in X$ in states of “failures” $\overline{x_i} = z_i$ and “non-failures” is x_i.
$q_i = (\ldots, x_{ik11}, x_{ik12}, \ldots, x_{ik21}, x_{ik22}, \ldots)$.

Here, $i = 0, 1, 2, \ldots, N_Q$; $N_Q = N_d - 1$, N_d is the number of binary bit codes, and $N_Q = N_d - 1$ is the number of states.

On the example of a system of three or five elements: Code tables can be drawn in the same way as in subsection 3.7.3 (Table 3.2a, b).

In (3.35): $\sum_0$ is a set of conditions for determining a risk situation in the S system; ξ is the symbol of the outcome randomness factor when states change and, obviously, when codes change in the state change processes $q_{j1} \to q_{j2}$ depending on the combinations of failures (χ_i (ξ) $= 1$) and non-failures (χ_i (ξ) $= 0$) of elements $x_i \in X$ [17, 24]; ω_ξ is an element of some probability space of type U from (2.11) or (3.1); T is the interval of observation of the state change process in the system.

Two simple systems were considered here: types of systems № 1, № 2: S_1—material, S_2—non-material.

The general view of the presented models is suitable for describing the system of interaction of subsystems with the SHEL interface based on the RRS SST tools when analyzing the HF influence on flight safety (see Chap. 5 below).

Table 3.2 (**a**) Codes and states, $n = 3$; (**b**) Codes and states, $n = 5$

(a)					(b)						
State q_i	x_1	x_2	x_3	Codes	State q_i	x_1	x_2	x_3	x_4	x_5	Codes
q_0	0	0	0	d_1	q_0	0	0	0	0	0	
q_1	1	0	0	d_2	q_1	1	0	0	0	0	
q_2	0	1	0	d_3	q_2	0	1	0	0	0	
q_3	0	0	1	d_4	q_3	0	0	1	0	0	
q_4	1	1	0	d_5	.	…	…	…			
q_5	1	0	1	d_6	.	…	…	…			
q_6	0	1	1	d_7	.	1	1	0	0	0	
q_7	1	1	1	d_8	q_{31}	0	1	1	0	0	

3.12.2 Functional Worthiness and Risks of Accident Occurrence in ATSs

The following designations for the basic provisions of functional worthiness (FW) and functional probabilities (FP) of S_{FS} systems are adopted [20, 38, 48, 49].

It is demonstrated that with the help of the SST tools, it is possible to provide a formalized description of the technology and procedures in the maintenance and repair systems, since the operation of foreign-made aircraft involves the MEL program, where units and products are replaced based on the results of calculating risks by the criterion of reducing the level of risks and maintaining the level of safety. (Examples are given in Chap. 5.)

The starting position is the ICAO K_{FW} criterion.

The functional worthiness criterion $K_{\text{FW}*}$ is a given probability of the first failure τ_{FS} outside the guaranteed operational time interval T_*, i.e. (the probability of non-failure up to $t \geq T_*$:

$$K_{\text{FW}} = P\{t_{\text{FS}} \geq T_*\}|[t_0,T],\ \Sigma_0,t_{\text{FS}} \geq T_*,\ t_{\text{FS}} \sim A_{\text{FS}},\ t_{\text{FS}} \geq T_*,\bar{T} \in [t_0,T], \quad (3.36)$$

Here, A_{FS} is a functional failure (according to I. Aronov, M. Neymark) that can be written in the form of *FS* (failure state), according to Chap. 2, as:

$$A_{\text{FS}} = \{a_{ij}(\tau_{\text{FS}ij})\} \in \Omega_{\text{FA}} \cup \Omega_{\text{FS}} \equiv \Omega_{\Sigma},$$

$$\tau_{\text{FS}} = \text{S}_{\text{UP}}\big\{t_{ij} | \{t_{ij}\} \Rightarrow A_{\text{FS}} \neq \varnothing | \Sigma_0\big\}, \quad (3.37)$$

where a_{ij} is a failure of the elementary link of the S_{FW} system; T_* is the right boundary of the guaranteed average time to the first failure of $T_* \leq T$; T is the average operating time to one failure; t_0 is the zero-time reference of observation (test).

On this basis, the functional *failure* A_{FS}, according to (3.37), is the *disjunction of conjunctions* of failures of elementary links with damage in the critical state $\tilde{A}_{FS}$ (according to I. Ryabinin). Each conjunction of the elements $\{a_{ij}\}$ is included in the chain of events leading to damage), i.e., leading to a catastrophe.

Now, it is not difficult to demonstrate that the probabilistic models of failures do not give anything new. Moreover, educational and scientific publications provide simplified (educational) models of systems that are not credible. However, in this form of describing the nature of phenomena, it is easy to apply the method of searching for event chains leading to a catastrophe and to estimate the risks by formulas (3.2), (3.3), i.e., within the RRS framework. It is so simple and natural that one can hope for the successful promotion of the ideas of the book in the practice of flight operations in civil aviation.

Examples are given in Chap. 5 (below).

The basic provisions of safety management based on the models considered are as follows.

Provision 1-FW: K_{FA} and $\hat{R}$ in various regions of FW are parameter.

Provision 2-FW: An acceptable risk $\hat{R}$ for the estimate $\tilde{R}_*$ (according to G. Gipich) will be:

(a) for damage—scalar $\tilde{H}_{R*}$, $\hat{H}_{R*}$, to ensure $\bar{A}_{FS}\,(t_0, T_*)$;
(b) for a set of indicators $\{\tilde{R}_*, \tilde{H}_{R*}, \hat{H}_{R*}, \tilde{\mu}_{1*}\} = (t_0, T_*) \sim \{1\text{ h}, 2\text{ h}, \ldots, T_*\} \Leftrightarrow A_{FS}\,(t_0, T_*)$.

Provision 3-FW: Flight safety management in the case of *FS*:

(a) through the management with risk $\hat{R}$ of events, processes, and its estimates $\tilde{R}$ for the found *R*;
(b) risk event chain management in the *ESTOP* system—in the S_{FW} structure;
(c) FW *redundancy* (in terms of flight safety)—*no more than duplication*;
(d) management—by resetting (in time) of the S_{FAO};
(e) recovery by K_{FA} (according to Cox, Smith) (state control).

3.12.3 Classification of Risk Events in the Space of Discrete States

Classification is created to build a universal set of ATS attributes required for the analysis of event scenarios and risk assessment using the Fuzzy Sets methodology. The proposed classification refers to systems with a discrete space of states *Q* from Sect. 3.12.2.

For *binary partitions*, the following relations are valid:

$$\Omega_\Sigma = \Omega_0 \cup \Omega_R \cup \varnothing : \Omega_0 = \{A_\xi = \bar{R} \neq R\},\ \Omega_R = \left\{A_\xi = A_\xi^* = R\right\}, \qquad (3.38)$$

where $\varnothing$ denotes an empty set, $\bar{R}$ is an event of type "no risk" or "chance", Ω_R can be considered a conditional space of risk events; $\Omega_R \in \Omega_\Sigma$. In the general case, this will be as follows:

$$R = R(\xi|\Sigma_0, H_\xi);\ \bar{R} = R(\xi|\Sigma_0, H_\xi); \bar{R} \cap R = \bar{\varnothing}. \tag{3.39}$$

Events are considered risky, if each outcome is associated with negative or undesirable consequences ("loss") in some experiments.

Simple risks are defined in the conditional risk space Ω_R and identified with a space $Q_\varphi\{\varphi\}$ of factors $\Omega_\varphi\{\varphi_i\}$:

$$\Omega_\varphi = \cup\varphi_i : \cap\varphi_i = \varnothing \text{ with } R_i \Leftrightarrow \varphi_{I(j)}. \tag{3.40}$$

The model provides the simplest options for assessing (simple) risks in the form of $\tilde{R}$ based (3.40) on a two-dimensional set of indicators.

With the help of the safety theory provisions, objective conditions for the occurrence of a catastrophe can be revealed in the form of a system error or a logical combination of the elements $x_i \in X$ as a possible path leading to a catastrophe.

It is necessary to determine all the measures of risk and μ_{R1} introduced in (3.2); and, first of all, μ_{R2}; to take into account the traditional probabilistic measures for μ_{R1} on the basis of the S system properties. Further, it is possible to consider both simple R_i and composed risks in the sense of risk events, as suggested in the RRS in Chap. 2.

The significance of the integral risk value is assessed according to (3.3) on the basis of the integral risk indicator $\hat{R}$.

3.13 Classifier of Risk Event Uncertainty Types

3.13.1 Definitions in the Uncertainty of Risk Events

In the period from 1990 to the present, the concept of system safety in the global aviation community has shifted from the definition of a simple symbol and "spell", especially in aviation, to a scientific category due to the seriousness of the "rare events" problem that occur with the «near-zero» probability (per ICAO and NASA). It was recognized that the causes of catastrophes are "*not probabilities*" (as sometimes stated in the PSA and in the RT), but "*system errors*" that can only be estimated on the basis of models for calculating risks of negative events determined on the "tails" of the probability density distributions for off-design rare events, on which there is no reliable statistics [34].

The cause of accidents is a chain of events or a scenario with the system entering into a hazardous state, through J. Reason chains (per ICAO), the probability of which is completely irrelevant if the damage from the accident is significant and unacceptable for system users [34]. Thus, the problem in the safety theory was reduced to the search for new programs to ensure flight safety (and industrial safety). Necessary solutions are provided within a new doctrine "*Reliability, Risk, Safety*" formulated in the framework of the system safety theory [12, 26].

Based on the known recommendations by ICAO, the default postulate is as follows: "***Reliability is the basis of safety***, but with the provisions for "reliability" only ***safety cannot be assessed*** and, even more so, it is not possible to effectively ensure "safety". It is ***incorrect*** to introduce, by analogy with reliability, the "***average time to catastrophe***" ***in case of rare events,*** **etc**. The methods for hazard (safety) assessing by additional sources of information are unreliable without statistics [34].

It is really impossible and unprofitable to provide absolute reliability by mathematical and technical means, since any increase in reliability is associated with production costs, which should be, to a certain extent, optimally based on justified criteria of optimality and efficiency [26].

System safety can be ensured within acceptable requirements only by reducing the risks of catastrophes on the basis of risk management methods taking into account the risk factors in accordance with the system safety theory methods [26, 34] and using safety management systems (SMSs).

Based on [12, 26], an SMS is defined as follows: "A safety management system is a set of interrelated and ordered elements and modules of the sets type (in the minimum composition, per Annex 19) designed to achieve the management goal of ensuring the necessary level of flight safety in accordance with the adopted system approach" (according to the British Standard) [26].

3.13.2 Types of Information Uncertainty in SMSs

In the system safety theory, the following types of uncertainties can be identified (by functional characteristics), which provide a number of models [26, 46]: deterministic; statistically determined (clear, statistically deterministic); fuzzy models on fuzzy subsets of objects; game models and "fractal" processes.

This measure is not random. It determines the probability of an event in a probabilistic sense *for rise event (**R**) and chance event (**B**)*, etc.

Uncertainty of the "randomness" type reflects the *property of measurability of functions from a random event as* a set of clear probability density functions and the existence of a mathematical expectation, variance, with reliable statistics.

The "*probabilistic approach*" led to misperceptions in the mathematical sense and to absurd results in the "rare events" problem when determining the level and significance of risk through the concept of "average risk" modeled on the financial domain and the reliability theory for technical systems [26]. It was noted above that a random variable is a parameter or quantity the value of which cannot be predicted in advance, but its probabilistic properties are deterministic and clear.

So, if these properties do not exist, any appropriate computations are impossible in principle.

Therefore, the indicator: “***PROBABILITY OF CATASTROPHE NON-OCCURRENCE” is completely EXCLUDED from consideration in this book. (The opinion of the authors of the book is that it is necessary to completely revise a number of known standards in order to exclude contradictory opinions on the issue under consideration.)***

From this, it follows: “IT IS UNACCEPTABLE TO REPLACE THE FUZZY CONCEPT OF “POSSIBILITY” with the word “PROBABILITY”, which is not objectively found as a number (clearly) in situations with rare events” (i.e., it is possible, if there is no “rare events problem”). In the system safety theory, the following types of uncertainties can be identified (by characteristics), which provide a number of models: deterministic; statistically determined (clear, statistically deterministic); fuzzy models on fuzzy subsets of objects; game models and “fractal” processes.

“*Chance is a fuzzy measure of the amount of “luck”* in the experience or in the system state under conditions of a favorable predictive event ***B*** (*chance*)”.

Only any indicative estimates of the flight safety level in the STS state are determined as [0, 1, D_R] for admissible STS states on the basis of the DB:

0—*safe,* 1—*hazardous, acceptable* (with fuzzy indicators like “*not very*”, “*reasonably*”, “*sufficiently”*, etc.).

Since the probabilistic approach does *not provide certainty.*

Different types of uncertainties are classified as follows:

(a) “*Randomness*”, but provided that the probabilities and probability density functions are clearly described and known. In this case, all necessary computations are performed using known probability density functions following the normal Gaussian law or others based on statistics [14].
(b) “*Fuzziness*” is an uncertainty type when all computations of uncertainty indicators are performed on the basis of the L. Zadeh function [11], through membership functions μ (mu), a “measure of the truth of the statement”.
(c) “*Minimax*” uncertainty estimated on the basis of calculations by the minimax method in problems and situations similar to “game uncertainty”, but with some information about the maximum achievable and known levels of restrictions on changes in uncertain parameters [50].
(d) “*Game*” uncertainty manifested in the games theory in the absence of a priori and a posteriori information about possible actions of players, which are discussed below in the subsequent subsections.
(e) “*Fractals*”.

The five groups (types) of uncertainties define a set of models depending on the degree of uncertainty of information about the parameters of models and differ by the following features:

- *deterministic model*;
- *statistical definite values and models*, which is also classified as “statistical determinacy” [21];
- *fuzzy models*, i.e., models on Fuzzy Sets of objects;
- *game models*.

3.13.3 New Principles of Constructing SMSs in the Fuzzy Sets Class

The general principle of creating SMSs is based on the scenario approach to the assessment of flight safety in civil aviation and in the transition to fuzzy subsets of SMS attributes using the Fuzzy Sets methodology [26].

The problem is that it is *necessary to assess the integral levels of risks* and chances in the *"rare events"* problem in the absence of reliable statistics on the measures of catastrophe possibilities.

The SMS functional scheme and structure are defined, as well as the list and functions of the main modules of the core of this system based on the ICAO recommendations [34]. Based on the ***Fuzzy Sets*** and the adopted classifier of uncertainty and unpredictability of "rare events", appropriate procedures are developed for assessing safety indicators and levels of risks of negative events, taking into account the accumulated database on the significance levels of the predicted adverse factors. At the same time, preventive management actions are determined that affect the system state and make it possible to exclude the occurrence of accidents or catastrophes (or reduce the risks of their occurrence) in advance before they can occur due to the manifestation of factors under a known set of threats. The main idea here is the application of rules and methods for weighting *risks and chances* in making decisions on managing the state of systems [26, 46].

In SMSs, according to the proposed definition [26], for situations with rare events it is necessary to adopt following applications:

$$\begin{aligned}&\text{"}\mathit{Risk}\text{ is a fuzzy measure of the amount of hazard in STS states}\\&\quad\text{when a threat and hazardous factors are }\mathit{detected}\text{" so as follows:}\\&\quad\text{(risk is "}\mathit{great}\text{", "}\mathit{small}\text{", "}\mathit{acceptable}\text{").}\end{aligned} \tag{3.41}$$

According to the "Oxford glossary", "risk is a possibility of occurrences (harm, damage)" upon the hazards or threats in any prognosis conditions.

Similarly, "chance is a fuzzy measure of the amount of "luck" in the experience or in the system state under conditions of a favorable predictive event **B** (*chance*).

Only indicator estimates of the flight safety level in the STS state are determined as $[0, 1, D_R]$ for admissible STS states on the basis of the DB:

0—*safe*, 1—*hazardous, acceptable* (with fuzzy indicators like *"not very"*, *"reasonably"*, *"sufficiently"*, etc.).

Since the probabilistic approach does not provide certainty.

Note. Value D_R is a fuzzy estimate [26] of the permissible STS state on the basis of the DB as a result of the "fuzzy implication" procedure in the form of ***"Inputs"* → *"Outputs"***.

3.13.4 General Scheme of Risk Identification in SMSs (with Fuzzy Sets)

The system safety theory (SST) is determined through uncertainty indicators from the classifier by the ***Fuzzy Sets*** methodology. The concept of "risk is a probability…" with the "near-zero" probability of rare events is completely abandoned, since this idea is impossible to implement [12, 26, 34].

The proactive mitigation process per ICAO (NASA) in the SMS is implemented using corrective actions based on the corresponding procedures per the following algorithm [26]:

$$\text{Threat-risk event-prediction of the scenario of events leading to a catastrophe -hazardous state-risk assessment-management action.} \tag{3.42}$$

The proactive management solutions are found based on event categories such as when the ICAO "*processability*" is abandoned. The significance of risks is suggested to be evaluated on the basis of a two-dimensional risk assessment model (E. Kuklev's formulas, an analog of the ICAO concept, but in mathematical form [12, 26, 34, 46]). Here, the following relationships have being assumed for summary:

$$\tilde{R} \to \tilde{R}_* \Rightarrow \tilde{R}_* = \left\{\tilde{R}_{*j}\right\},$$

$$\hat{R} = f_R\left(\tilde{R}_*|\Sigma_0\right) \equiv f_R\left(\mu_{1*}, \mu_{2*}, \tilde{H}_*|\Sigma_0\right),$$

The following indicators of truth are used in form μ_1 as a measure of risk of the first kind, indicating the uncertainty (or randomness) of occurrence for a risk event R with a negative result $\tilde{H}_R$. Symbol $\tilde{H}_R$ is a measure of the consequences or damage. It is the cost of risk or the "severity" of damage. Symbol $\sum_0$ is describing the experience conditions or parameters of the situation in the operation of the system (hazard class and model of the system hazard, type of the "scenario", the event tree based on *FMEA*, but in the form of "Xmas tree" [26]. The state changing on graph of states of catastrophic system failures images the specified for the method of minimal cut sets of failures [46]. Value $\hat{\tilde{R}}$ is an integral risk (with fuzzy estimates according to [12], i.e., the amount of hazard in the formulas (3.8) in a given state of the system [46].

3.13.5 Weighting Risks and Chances

"Optimal solutions" for proactive management of the system state are taken by weighing "risks" and "chances" for predicted events in conditional binary spaces of outcomes in fuzzy subsets when assessing the significance of integral risks—according to ICAO (NASA) matrices.

The risk analysis matrix (ICAO) is used on the basis of the methodological provisions of the Fuzzy Sets (and subsets) theory. Here, interpretations of the introduced concepts are given within the framework of concepts for Fuzzy Sets by M. Fujita (Tails—far from medium [23]) and G. Malinetskiy ("Hard Tails" from the Risk Theory—RAS ICS).

The scientific problem in (1), (2) is the construction of functions for estimating the quality (3) of a set of elements in (2) that, in the general case, are not part of vector spaces that are dense everywhere. In the general case (fuzzy estimates, etc.), it is inadmissible to normalize sets (2), (3) to a scalar vector convolution, since elements in (2) do not form any topological space [26].

In summary the decision must be declared that the *Fuzzy Sets* scientific approach is a unique and the only method for investigation of parameters of different events multitude. The substantiation of the objective fuzziness of ICAO (NASA) matrices was given earlier by group of authors (G. Gipich, V. Evdokimov, E. Kuklev, M. Smurov), as a presentation in Montreal and in Washington (at the Boeing Company—period of 2012–2013).

For Fuzzy Sets, an example is given from American practice (and civil aviation in Russia) in the form of the task "On 3 PICs and the departure decision under the conditions of wing and body icing".

It may be noted that the applicability of the SMS concept for the medical practice of expert analysis based on the theory of risks has been found. The fundamental difference is that *in medicine, there is no well-formed mathematical theory and methodology for calculating risks* in full compliance *with the principles of* ***Fuzzy Sets***. However, in civil aviation of the Russian Federation, similar, but more general, tables have been developed on the basis of tables in medicine, which confirm the need for the development of SMS ideas and their wide implementation in the management of aviation activities in civil aviation of the Russian Federation, including the prevention of accidents and catastrophe conditions, falls of satellites such as "Phobos", losses of shuttles, catastrophes at hydraulic power stations, etc.

The implementation of SMS developments ***in the field of nuclear energy*** from civil aviation of the Russian Federation ***seems promising***. Examples of some developed risk tables are as follows: *table of railway track accident rate* (the catastrophe in the Moscow subway); medical risk map; payment risk matrix at the Nash point for a discrete minimax criterion for assessing safety of Cessna sports aircraft. ***The task in safety theory is the search for new principles and methods for ensuring flight safety*** (and industrial safety, taking into account the quality requirements to systems, such as "*Reliability, Safety*". One must accept, by default, the ICAO postulate that ***"Reliability is the basis of safety", but safety (in the sense of absence or prevention of possible harm) depends functionally on the system structure*** only and the conditions of its operation or practical application in off-design conditions that are rare, but can become cause unjustified consequences. ***So, it is completely incorrect*** to introduce, by analogy with reliability, the ***"average time to the catastrophe", etc. In the case of rare events***, methods for assessing a hazard (safety) for real and unstable statistics are not reliable. The safety of systems can be ensured within the framework of acceptable requirements only by reducing the risks of catastrophes on the basis of risk management methods, taking into account the uncertainty of occurrence of risk factors and safety threats in accordance with safety management system (SMS) application methods. In view of the special properties of rare events caused by the absence of a deterministic measure of the possibility of their occurrence and damage, which can be very significant in single catastrophic states, it is necessary to study the properties of phenomena with the "near-zero" possibility. This led, as was shown above, to the development of the so-called risk-oriented approach. But in this approach, it is necessary to study in more detail various types of uncertainty that are known at the present time and existence of these "uncertainties".

NASA has formulated an appropriate risk concept based on the notion of "fuzziness"—in the sense of the achievements of Berkeley's scientific school. At the same time, it was *found that the solution can be found on the basis of **Fuzzy Sets** methods* due to the difficulties in assessing system safety indicators for rare events. This position was adopted when constructing a special matrix "risk assessment" adopted by ICAO as a tool in a typical SMS. (Features of this approach are studied in detail in this book.)

This subsection provides a classification of uncertainty types based on the generalization of the NASA developments (12) and developments in a number of works of the Russian Academy of Sciences in the Russian Federation.

A *random variable* is a parameter or a physical quantity of the value which cannot be predicted in advance, but its probabilistic properties are deterministic and clear [26, 34].

The corresponding classifier is presented in Table 3.3. It is proposed to consider only five (or four) types of uncertainties as the main ones that are important in *OTSs* and *MSPs*.

3.13.6 Classifier of Information Uncertainty Types

In fuzzy (non-Gaussian) models, ***uncertainty is only the non-truth of statements*** about the values of the observed quantities. The ones can be estimated with the help of the proximity measure of any fuzzy value from the fuzzy subset from some universal set carrier of the true value. Such position according of risk concept for Zadeh [5] ***measure of truth for "risk"– is*** μ ***~ (muy)'*** [5].

At the same time, computational operations and logical transformations of complex fuzzy operations of type (1) with the description of phenomena properties on the basis of *"linguistic" terms and fuzzy logical relationships such as "the risk is great", "the chance is really significant"*, etc, can be equally used based on the fuzzy logic methods.

With this approach, it is possible to overcome the problem of "extracting information from the "«zero»", arising in the PSA method in solving problems with rare events ("near-zero" probability).

Table 3.3 Classification of uncertainty

No.	Features γ_i	Uncertainty types	Designation of conditions Σ_{0i}
1	γ_1	***"Randomness"****—the probabilities and probability density functions are clearly described and known. Computations are performed precisely using known probability density functions or based on statistics*	Σ_{01}
2	γ_2	***"Fuzziness"*** is an uncertainty type, the measure of which is estimated on the basis of the L. Zadeh function [11] through membership functions μ (mu).	Σ_{02}
3	γ_3	**"Minimax" uncertainty** estimated by the minimax method with some information about the maximum achievable and known levels or intervals of restrictions on changes in uncertain parameters	Σ_{03}
4	γ_4	***"Game" uncertainty*** manifested in the games theory in the absence of a priori and a posteriori information about possible actions of players	Σ_{04}
5	γ_5	"Fractal's processes"	Σ_{05}

The materials presented here are necessary for the development of "risk models" regarding situations with rare events specified in Chaps. 4 and 5.

Summary from item 3.13.6. The main recommendation is as follows: "It is unacceptable to replace the fuzzy *CONCEPT of "possibility"* with the word *"PROBABILITY"*, which is not objectively found as a number (*clearly*) in situations with rare events" (i.e., it is possible, if there *is no "rare events problem"*).

A random variable is a parameter or quantity the value of which cannot be predicted in advance, as noted above, in a particular experience, but its probabilistic properties are deterministic and clear.

Thus, in the simulation of random processes, the generalized characteristics of these processes are used, known from probability theory and mathematical statistics. Among them (and above all), clear (calculated) indicators such as mathematical expectation and variances of quantities, probability density function (PDF), and cumulative distribution function (CDF) are used. Tools for such modeling are known and described in [34].

Probability, according to classical works [21, 34], is a *measure of randomness* of event occurrence. But *this measure is not random but clear*, which clearly defines the possibility of event occurrence in a probabilistic sense. Formally, this is a numerical value of the clear quantity of the probability measured (or found) in the range from 0 to 1 on the basis of provisions and theorems of the probability theory (or statistically if the rule of "law of large numbers" and correct restrictions are observed). The main (determining) restriction for the use of the uncertainty concepts of the "randomness" type in the SMS is the property of measurability of a random event and a *reliable statistical database or a* ***set of analytical (clear) density functions***.

3.13.7 Definitions and Principles of Constructing SMSs Based on Risk Calculation Models

Background ICAO ("Annex 19") announced the program for creating SMSs on the following basis [26]: to provide SMS requirements (ARMS-ECAST type), but with the solution of the "rare events" problem (per ICAO) by creating databases on NASA samples (*principles and approaches*), taking into account the limited volume of statistics and with the uncertainty of the possibility measure (or *randomness, in simple situations*) for the occurrence of rare events; it is recommended to apply "Preventive" (proactive) management of the STS state taking into account risk factors on the basis of ICAO algorithms with limited statistics (with uncertainty of the measure of rare event occurrence); ensure [34] "Monitoring of the flight safety state".

It is established that the solution can be found on the basis of the ***Fuzzy Sets*** *method* due to the difficulties in assessing system safety indicators for rare events. This leads to the need to create special safety management systems, such as SMSs, as shown above.

The main terming so as "risk" is an symbol for uncertainty of danger with fuzzy prognoses results (harm) and unexpected appearance of rare risk event (probability "almost zero") must be accepted for refuse from "risk probability". It must be accepted also that so rare phenomena as the risk event R is very rare. Nancy Levisohn (from Netherlands) [47] had noticed that "application of probabilistic concept "of risk in the theory of safety" incorrectly in situations with rare events.

Thus, according to a measure of danger (on G. Malinevskiy [30]) for scenarios resulting in the points "of vulnerability" and "windows of vulnerability" as rare events in [34] is need to confirm the "*new* classification of intercommunication" of basic terms corresponds it in the theory of safety of the systems (safety, danger, threat, vulnerability etc.). Then, the formula of estimation of integral size of risk must be entered so as it was proposed above at the publication.

Mathematical description of risk concept on ICAO

In official, description is presented here as well-known and unclear kind for [13–15]:

Risk Concept: Likelihood (L) and Severity of Harm (H) (**)

And more &-logical symbol of combination operators for any elements of sets.

But here, in ***(**)***, the methods of measuring of risk and types of combination of meaningful parameters in this conception are not indicated.

There are only the general pointing on factors *(L)* and *(H), and there is no* risk meaningfulness. Generally, unclear estimations of risk are accepted here without the certain meaningfulness which also not indicated.

However, the one must be done.

For example, based on fuzzy the implication from item 3.1.1 by using (3.1):

$$T : V \to [0{,}1] \Rightarrow T\,(P \supset Q) \to [0{,}1)$$

$$T(P \supset Q) = \max \{min\ \{T(P), T(Q)\}, 1 - T(P)\}$$

where P and Q are the simple recital suggestion in combination(P or Q), ($P < 0$), etc. according to operation & in (**) (above).

The methods of evaluation of risk levels must be considered in a next unclear view (classic—*by L. Zadeh* from [16, 51]):

(**risk is large**, **risk small**, **risk is not meaningful**, **etc**.)

Model of risk event as a rare phenomenon

Similar risk event with uncertainty (unclear) of negative consequences [5] so as harm from risk event and serious unclearness of possibility of appearance is presented here in the next forms as follows [31]:

$$R = R(\gamma | Z_R,\ \mu,\ H_R,\ \Sigma_0), \tag{3.43}$$

where Z_R is a vagueness of the set kind γ from a classifier; μ is an unclear measure of possibility of origin of risk event, for example, as a loss of property of functionality (i.e., "failures" on FMEA [5, 16]; H_R is a damage; Σ_0 are terms of origin of the near accident (scenario) conditioned by the threat of Z_R; a sign **(... |...)** is a line in (3.43) that designates properties of experiment, giving the event of R entered in consideration taking into account terms characterizing the set experiments.

Problems are that some danger factors, included in parameters *as* (L) and (H) above *and also* (μ, H_R), are not be created in topological spaces.

It is more succeeded to find the mathematical formula of danger of situation according to the conception of risk (and danger) situation (3.44) in kind of $\tilde{R}$

[16, 31] by ICAO in *(**)* [5].

Then, the danger situation $\mathrm{N} = 2^{\mathrm{n}}$ *will be done as the cortege*:

$$\tilde{R} = \langle Z_R, \mu, H_R | \Sigma_0 \rangle, \tag{3.44}$$

where μ (or μ_1) is a vagueness of some kind from a classifier [46] so that is an unclear measure of danger possibility of origin of risk event, for example, as "a loss of property of functionality" on [51] or it is a damage [10, 30].

Thus, it may to get the algorithm for level evaluation of integral risk by $i = \overline{1,\ m}$, based on monitoring of hazardous and danger factors in the systems of type of safety management system (SMS) [12] (in accordance with matrix of NASA) as:

$$\hat{\tilde{R}} = \hat{f}\left(\tilde{R}|\Sigma_0\right) = \hat{f}(Z_R,\ \mu,\ \mu_1,\ H_R\Sigma_0), \tag{3.45}$$

where $\hat{\tilde{R}}$ is total level of risk on fuzzy interpretation (3.45) by NASA based on logical symbol "&" in *(**).*

Thus, the risk's situation (3.44) by ICAO determined in the kind (3.45) into concept of "*danger*" in the system through the hazard for safety of Z_R which must be accepted in all well-known theories [13–15]. Then, the unclear *("verbal")* formula (above) of ICAO will be considered at the kind that allows to find the size of integral risk $\hat{\tilde{R}}$ (*or in kind* $\tilde{R}$) by *NASA*.

The hazard (or threat) in view Z_R plugs may be determinate as the sets of factors dangers that are results of origin hazardous risk occurrence of alternative chains of events leading to the catastrophe in the corresponding critical scenario of events.

The estimation of scenario danger $\sigma^2 - 1$ (*or in kind* $1/\lambda$) proposed in (3.45) with an account with the appearance of elements from cortege with two arguments (minimum) with pointing of kind or class of threat of Z_R with some sets of dangerous factors of $\phi_i \in Z_R$:

$$\varphi_i \in \Phi = \{\varphi_{\rm i}|Z_R,\ \Sigma_0\}. \tag{3.46}$$

It is the basis of indicators for risk levels, finding by means of matrix, accepted earlier. It provides the solving of problem of rare events. An amount or measure of danger is determined here in the view as it is a function from multitude with two arguments so as $\hat{\tilde{R}}$ (or $\tilde{R}$) cortege (3.44) accepted above in description of $\hat{\tilde{R}}$. Critical failures so as catastrophes in systems may be found upon the method of minimal sets of failures by means of use a tree of events by *FMEA* [46] or *"sigma-algebra" E.*

3.14 Conclusions

1. It is advisable to develop methods for assessing system safety with transition to the theory of calculating non-trivial risks conditioned by the possibilities of the occurrence of rare events, and especially those that are represented in the form of J. Reason chains. In the RT, the use of the PSA method provides proper solutions mainly for problems of estimating trivial risks with reliable statistics.
2. The *transition* should be *based on the application of the concept of fuzzy subsets* and non-stochastic modeling of *discrete state change processes in systems in the form of J. Reason chains*, which are based on the concepts of MCF, FMEA, and CATS in the classical RT, but are applicable in solving rare events problems. This

ensures the harmonization of various approaches to solving rare events problems using the Fuzzy Sets approach.

References

1. Orlovskiy SA (1981) Problems of decision making with fuzzy source information. “Science” FM, Moscow (in Russian)
2. Rybin VV (2007) Fundamentals of the fuzzy sets theory and fuzzy logic—study guide. STU Moscow State Aviation Institute. Moscow, 95 p (in Russian)
3. Kuklev EA, Evdokimov VG Predicting the safety level for aviation systems based on risk models. J TR 2(45):51–53 (in Russian)
4. Kuklev EA (2005) Estimation of catastrophe risks in highly reliable systems. In: Materials of the 13th international conference “problems of complex system safety management”. RAS ICS. Moscow, 2005, 55 p (in Russian)
5. Probabilistic Risk Assessment Procedures for NASA Managers and Practitioners—Office of Safety and Mission Assurance NASA. Washington, DC 20546, August 2002 (Version 2/2)
6. SMM: Doc. 9859-A/N460. ICAO, 2009
7. General rules for risk assessment and management (resource management at life cycle stages, risk and reliability analysis management—URRAN). JSC Russian Railways standard STO RR 1.02.034-2010. Moscow: 2010 (in Russian)
8. CATS (Casual Aviation Technical Systems) (2012) Simulation of cause-effect relations in aviation systems on the basis of risk assessment. Research of the Air Accident Investigation Commission (Netherlands). ICAO (in Russian)
9. Smurov MY, Kuklev EA, Evdokimov VG, Gipich GN (2012) Development of tools for assessing the risks of occurrence of risks of AUI in the AASS of the airport system. J Transp Russian Federation, 2(39), St. Petersburg (in Russian)
10. Documents of the 37th ICAO Assembly (October 2011, Montreal)
11. Malinetskiy GG, Kulba VV, Kosyachenko SA, Shnirman MG et al (2000) Risk management. Risk. Sustainable development. Synergetics. Moscow: Nauka, 431p. Series “Cybernetics”, RAS (in Russian)
12. SMS & B-RSA (2008): “Boeing”, 2012
13. SMM (Safety Management Manual): Doc 9859_AN474 – Doc FAA: 2012
14. Aronov IZ (1998) Modern problems of safety of technical systems and risk analysis. Stand Qual 3 (in Russian)
15. British Standart (1992) Quality management and quality-assurance. Vocabulary. VS EN ISO-8402
16. Volodin VV (ed) (1993) Reliability in technology. Scientific-technical, economic and legal aspects of reliability. Blagonravov Mechanical Engineering Research Institute, ISTC “Reliability of Machines”—RAS, Moscow: pp 119–123 (in Russian)
17. Prokhorov PZ, Rozanov YA (1987) Probability theory (basic concepts, limit theorems, random processes) “Science” FM, Moscow (in Russian)
18. Korolev VY, Bening VE Shorgin SY (2011) Mathematical foundations of the theory of risk. Fizmatlit, Moscow (in Russian)
19. Gipich GN (2005) The concept and models of predicting and mitigating risks in ensuring the airworthiness of civil aircraft. Moscow State University. TEIS, Moscow (in Russian)
20. Evdokimov VG, Komarova YV, Kuklev EA, Chinyuchin YM (2013) Quantitative estimation of the possibility of occurrence of accidents in aviation complexes. Sci Bull MSTU CA. No. 187(1):53–56 (in Russian)
21. Accident Prevention Manual. Doc. 9422-AN/923. International Civil Aviation Organization, 1984

22. Risk management—Principles and guidelines. Standards (Australia)—AS/NSZ ISO 31000: 2009
23. Fujita M (2009) Frequency of rare event occurrences (ICAO collision risk model for separation minima). RVSM. ICAO, Doc. 2458. Tokyo: EIWAC 2009
24. Ryabinin IA (1997) Reliability, survivability and safety of ship electric power systems. Kuznetsov Naval Academy, St. Petersburg (in Russian)
25. Relex Faut Tree Module (FMEA)—473.Relex.0709. St. Petersburg: 2011 (in Russian)
26. Aronov IZ et al (2009) Reliability and safety of technical systems. Moscow (in Russian)
27. Livanov VD, Novozhilov GV, Neymark MS (2013) Flight safety management system. IL SMS. "AviaSoyuz", 1:14–21 (in Russian)
28. Guidance on hazard identification. SMS WG (ECAST (ESSI) Working Group on Safety Management System and Safety Culture). Handwritten, 2010 (in Russian)
29. Orlov AI, Pugach OV (2011) Approaches to the general theory of risk. RFBR Grant-2010. Bauman MSTU. Moscow (in Russian)
30. Severtsev NA (ed) (2008) Kuklev EA Fundamentals of the system safety theory, RAS Dorodnitsin CC. Moscow, pp 175–180 (in Russian)
31. McCarthy J (1999) U.S. Naval Research Labaratory; Schwartz N., AT & T. Modeling Risk with the Flight Operations Risk Assessment System (FORAS). ICAO Conference in Rio de Janeiro, Brasil, Nov. 1999
32. Risk management—Vocabulary. ISO Guide 73: 2009 (E/F). BSI: 2009
33. Gipich G, Evdokimov V, Kuklev E, Mirzayanov F (2013) Tools: identification of hazard & assessment of risk. Report at the ICAO Meeting "Face to Face". "Boeing" Corp., Washington, janv.
34. Annex 19: C-WP / 13935 – ANC Report (March 2013), based on AN-WP / 8680 (Find) Review of the Air Navigation Commission, Montreal–Canada
35. Smurov MY, Kuklev EA, Evdokimov VG, Gipich GN (2012) Safety of civil aircraft flights taking into account the risks of negative events. J Transp Russian Federation 1(38):54–58 (in Russian)
36. Safety Management Systems—Guidance to Organization. (Vers-3). Safety Regulation Group, 2010
37. Krylov DA (1993) TPP, NPP: Hazard and risk. Energy 3 (in Russian)
38. Gipich GN, Evdokimov VG (2009) Principles of a unified approach to assessing complex system safety based on indicators of risks. In: Coll. of papers. RAS CC "system safety", No. 2, Moscow, 2009, pp 112–120 (in Russian)
39. Risk management—vocabulary—guidelines for use in standarts. PD ISO / IEC, Guide 73: 2002.B51: 2009
40. Bykov AA, Demin VF, Shevelyov YV (1989) Development of the basics of risk analysis and safety management // Collection of scientific papers of the Kurchatov Institute of Atomic Energy. Moscow: Publ. House IAE (in Russian)
41. Barzilovich EY, Kashtanov VA, Kovalenko IN (1971) On minimax criteria in reliability problems. ASUSSR Bulletin. Ser. "Technical Cybernetics", No. 3. Moscow, pp 87–98 (in Russian)
42. Henley EJ, Kumamoto H (1984) Reliability engineering and risk assessment. Mechanical Engineering, Moscow (Translated from English, Edited by Syromyatnikov VS) (in Russian)
43. Evdokimov VG (2010) Modern approaches to safety management based on the risk theory. International Aviation and Space Magazine "AviaSouz". NQ 5/6 (33) October–December (in Russian)
44. Evdokimov VG, Gipich GN (2006) Some Issues in the Methodology of Assessment and Prediction of Air Transport Risks. In: The 5th international conference "aviation and astronautic science 2006", abstracts, Moscow pp 75–76 (in Russian)
45. Documents of the ICAO High-level Conference (SMM). March 2010, Montreal
46. Amer M (2012) Younossy. 10 Things You Should Know About Safety Management Systems (SMS). SM ICG—Washington
47. Waitmann M UN: Japanese nuclear power plants are not ready for a tsunami. www.metronews.ru 02.06.11(in Russian)

48. Gipich GN, Evdokimov VG, Chinyuchin YM Basic provisions of the concept of building a safety management system for aviation activities. Sci Bull MSTU CA. No. 187(1):31–36 (in Russian)
49. Abramov BA, Gipich GN, Evdokimov VG, Chinyuchin YM Determination of the compliance service providers' safety management system to air transport standards. Sci Bull MSTU CA. No. 187(1):36–41 (in Russian)
50. Arnold VI (1995) Catastrophe theory. "Science" FM. Moscow (in Russian)
51. Accidents and catastrophes. Prevention and mitigation of consequences. Study guide. In 3 books. In: Kochetkov KE, Kotlyarevsky VA, Zabegayev AV (eds) Publ. House ASV, Moscow, 1995 (in Russian)

Chapter 4
Structure and Principles of Constructing the SMSs to Provide and Monitor System Safety Based on the RRS Risk Management Doctrine

In terms of the analysis of "risks", features of the design of special systems such as safety management systems (SMSs) are considered that are designed to manage such indicators of the system state as "safety" in the field of civil aviation. The creation of SMSs is conditioned by the need to transit to proactive methods of managing safety in the field of civil aviation, which is possible only on the basis of the methodology for calculating risks. In the flight safety domain, traditional (active–reactive) methods have already run its course, especially in the rare events problem.

This SMS was based on the recommendations of the new Annex 19 (ICAO, 2013) [1], which has a higher status than the SMM (Doc 9859-AN/460-2011) [2]. The NASA [3] and RRS doctrine (SST) recommendations and developments from Chaps. 2 and 3 of this book are also taken into account.

4.1 Standard International Requirements to the SMS Structure

4.1.1 Key Definitions and Purpose of the SMS

This book proposes the following definition:

An "safety management system" (SMS) is a set of interrelated and ordered elements or modules (in the minimum composition, per Annex 19) intended to achieve the management objective of providing the required level of flight safety in accordance with the adopted systematic approach (the alternative "Blue Folder" [4]—see below).

The ICAO systematic approach to flight safety management defines tools for the following: managing the state of systems; creation of a hierarchical structure of the SMS organization, including a module for the manager's responsibility for the flight safety in the business structure.

Kuklev E.A. et al., *Aviation System Risks and Safety*, Springer Aerospace Technology, https://doi.org/10.1007/978-981-13-8122-5_4

This definition is introduced in accordance with standards such as GOST R in the Russian Federation [5]. The content of the definition was discussed at ICAO workshops, as it was not in line with the primary version in Annex 19 (March 2013). However, it does not contradict the essence of the SMS and is more in line with scientific achievements in this field and the recommendations of the international British standard.

Note: The original definition recommended in Annex 19 is "*SMS is a systematic approach...*, etc."; the discrepancy between the meanings of "approach" and "system" has become the subject of discussion.

Types of SMSs by purpose and forms of declaration of functions are the following: SMS—flight safety management, AA SMS—safety management system for aviation activities.

Below (in Sect. 4.3), the possibility of creating an SMS is substantiated that is unified by functions (in the sense of the SST), but different in terms of the professional sphere of application in the form of 2 basic parts:

Part 1—"CORE"—a universal standard core for all SMSs;

Part 2—"DATABASES" (DB)—various, depending on providers.

SMSs (SMS–AA SMS) are designed for proactive and predictive safety management in aviation transport systems. The essence of SMSs in question is the identification and elimination of potential threats in the management of aircraft, finance, and investments in order to achieve such target results as ensuring flight safety by standard indicators, gaining real financial and economic benefits, for example, in the airline's activities in traditional or intermodal transportation. Such SMSs were developed in the "Boeing" corporation [6], in FAA [7, 8], in OJSC "Aeroflot". It is assumed that in highly reliable systems, risk events R are rare (with probabilities below 10^{-6} in Prdf) and have the "near-zero" probability), since other values cannot be determined exactly. This justifies the reasonable avoidance of the interpretations of "risk" as "probability", since events of this kind are detected through the values of their parameters only in pdf "tails" and cannot be described reliably. Pdf values for such parameters are very small.

There is no practical meaning in the multiplication of such unreliable quantities with the values of 10^{-7}, 10^{-8}, …, 10^{-12} used by the researchers in the PSA method, since the results obtained are not useful.

4.1.2 Integrated "SMS–QMS" Modules ("Blue Folder")

An SMS [4] is an active system integrated with other management systems to form a flexible technical and regulatory framework of the organization. At the same time, it identifies hazards and risks that are of importance to the entire organization.

In addition, at the same time, risk control is established in full compliance with reliability standards. Such an SMS focuses on hazards, and at the same time, due to the non-compliance with the strategic goals of the business organization (e.g., the "state/nation"), two mutually dependent QMS and SMS systems can affect safety [4].

Fig. 4.1 The front page of the paper prepared by the FAA ("10 Things You Should Know About Safety Management Systems (SMS)")

Therefore, it is necessary to develop an SMS integrated with the QMS. Generalized recommendations are given in the work ("Blue Folder") by A. Younossi (from the FAA) in Fig. 4.1. These provisions were presented in [9].

4.1.3 Main SMS Functions Recommended in Annex 19

Below are the main NASA provisions (page 219 in [3]):

– *identification of accident scenarios;*
– *estimation of the Likelihood of each scenario;*
– *evaluation of the consequences of each scenario;*

However, the term "Likelihood" should be considered (according to OXFORD) [1] as the "Possibility" of "Risk Events" with "near-zero probabilities", i.e., without the use of the term "probability", as accepted in the RRS doctrine (Chap. 2 of this book) and indicated in the SST.

From this, it follows that the key processes for the SMS are the following:

- identification of hazards associated with the activities of the organization;
- risk management based on a standard approach to risk assessment and control;

- predictive compensation of possible consequences in each (proactively) "detected" hazardous scenario of the development of the situation or a possible chain of events.

The NASA recommendation [3] was the basis of methods for proactive system safety management by the criterion "target safety level" using the risk value as an adjustable ATS output parameter.

The management triad known in ICAO (NASA) (Sect. 4.2.1 below) makes it possible to solve these issues.

4.2 Prediction of the Safety Level in the SMS for Complex Aviation Systems Using Risk Models for Critical Functional Failures

This section describes theoretical features of the methods used to assess levels and assure the safety of activities and operations in civil aviation on the basis of risk models.

4.2.1 Triad of Management Actions in the SMS

This triad (according to NASA [3]) comprises three types of ATS state management in current situation for $t \in [t_0, T)$ and in predicted or anticipated situations for $t \in [t_x, T)$ [5, 10].

The moment t_x is assigned to monitor the ATS state in the production activity, when it is necessary to ensure a guaranteed result in the planned operations, i.e., "in the future", t_0 is the initial moment of time for management and T is the planned period of ATS operation.

The types of safety state management per ICAO are those that were discussed in Chaps. 2 and 3 are listed as:

- active (reactive) (№ 1);
- proactive (№ 2);
- predictive (based on some a priori information (№ 3).

For each type of management, ICAO has developed guidelines recognized by the global aviation community. For each of them, NASA presents a general algorithm for constructing a management procedure, which is recommended for use in SMSs from various providers. The description is given in Sect. 4.2.4 as a scheme for developing scenarios to analyze hazardous situations in ATSs within the SMS.

The risk models adopted by the SST (RRS) on the basis of the *Fuzzy Sets* approach allow solving the tasks of this subsection using the positions of the general theory of managed systems. Here, an analogy with the terminal control can be seen according to [11].

From the glossary of the American Oxford University [1, 9], it follows that "*Risk is a possibility of serious (negative) consequences in the assumed situations under the conditions of determining threats of a given type so as: Threat is a source of hazard*".

This is the rationale for the RRS SST concept (Chap. 2) that risk is a *projected "amount of some hazard in a given state* of the system". Potentially, only *some fuzzy possibility* of occurrence of rare risk events can exist in the ATS, but they are *immeasurable (probabilities are unknown)*. The "probability" as a *clear quantity* can (and must) be *clearly (deterministically) calculated* only from known probability distribution functions (Prdf) and probability densities (pfd).

Management, for example, according to [11] (S. Kabanov), is a deliberate action (in time, space, including terminal control) that affects the selected facility or system, taking into account the measurement of the "discrepancy" (residual) of the objective function and the measured value.

In safety assurance and monitoring systems, it is possible to *reliably measure* (a priori—on the basis of models or a posteriori—on the basis of statistics) only *deviations of the values of control parameters* from standard ones. Otherwise, the concept of *"safety" is a "symbol"*, at most. The use of the concept of risk in the RRS as an "amount of hazard" (hazard measures according to the RAS) makes it possible to "measure risk" and its changes in a convenient physical form (measure), taking into account the variety of various adverse factors.

Indeed, one can take formula (3.2) from Chap. 3 to evaluate the significance of the risk $\tilde{P}$ and its integral value $\hat{P}$:

$$\tilde{R} = (\mu_1, H_R|\ \Sigma_0) \tag{4.1}$$

$$\hat{R} = f_R(\tilde{R}|\Sigma_0). \tag{4.2}$$

Any of the quantities—elements in (4.1) and (4.2)—can be fixed and included in a set of conditions Σ_0. Then one can obtain what was proved in the RRS: An indicator of risk (significance) can be any "clearly" or "fuzzily" measurable indicator:

$$\overline{R} \to \widetilde{R}_1 = \left(\mu_1, H_R|\sum_0\right) \Rightarrow \widetilde{R}_1 \to \underset{\sim}{\widehat{R}_1} \Rightarrow \text{chances with}\mu_1 \tag{4.3}$$

$$\tilde{R} \to \hat{R}_2 = f_R(H_R|\ \Sigma_0\mu_1) \Rightarrow \underset{\sim}{\hat{R}} \equiv \underset{\sim}{H_R} \to \text{chances with}\mu_1 \tag{4.4}$$

The values of (4.3) and (4.4) depend on the parameters of the systems and their sensitivity functions, which can be expressed through risk factors and parameters of the systems under study. This is almost impossible in the PSA (as "risk is a probability"), even with parametric variances and mathematical expectations in Prdf and pdf:

$$\sigma_x = f_\sigma(\{k_i\}), m_x = f_m(\{k_i\}) \ , \tag{4.5}$$

$$P_A(m_x, \sigma_x | \sum_0) \sim 10^{-6} \tag{4.6}$$

Therefore, according to the ICAO principles [1, 6], the state control in classical dynamic systems for civil aviation facilities should be performed using a certain controlled variable [11]. In cases of problems with "risk", one must consider a priori only predicted values.

In this sense, "risk management" is in line with the formulation of problems with "terminal control" [11] in the theory of optimal systems.

It is assumed that the measured (proactively, predictively) value is the risk $\hat{R}$ (integral characteristic of hazard). This value is compared with an *acceptable level of risk* $\hat{R}_*$, and a residual is determined (a risk defect $\Delta\hat{R} = \hat{R}_* - \hat{R}$.). This allows for the management of the system state taking into account the significance of the risk. The control (corrective) impact on systems depending on the residual $\Delta\hat{R}$ makes it possible to *"mitigate risks"* or *"accept them"* or *eliminate the possibility* (not the "probability") of *occurrence* of a predictable (*in advance*) risk event R in the system—before this event can occur. (The management method with the $\Delta\hat{R}$ value is given below in 4.4).

There are known SMSs built on the principles of risk management [1, 6] in "Boeing", "Airbus" corporations, and other companies, where the key point is the *prediction* of the occurrence of *"accidents and catastrophes designed in the system"* due to the objective existence of conditions for the occurrence of extremely "rare events—with the 'near-zero' probability, but causing great damage".

With this in mind, approaches [1, 6, 7] were proposed for risk assessment from (4.2) to (4.5) based on [7, 12, 13] for flight safety management too. This is also recommended for the management of safe activities of service providers, including the production and industrial manufacture of products such as airplanes, helicopters, engines, propellers [1, 7]: reactive (*i.e., immediate reaction* to the accident); proactive and predictive (predictable–proactive management of the system state by risk factors, as noted above).

Here, the required estimates from (4.3) to (4.5) can be calculated in two forms—through indicators and a matrix of risks with an estimate of the *residual risk* level in the form of "residuals":

$$\begin{aligned} \Delta I_{\hat{R}12} &= I_{\hat{R}2} - I_{\hat{R}1} \\ \Delta S_{\hat{R}12} &= S_{\hat{R}2} - S_{\hat{R}1} \, . \end{aligned} \tag{4.6}$$

In this *"triad" of management actions*, the key point is to take proactive measures to change the state of the system before the predicted hazardous event R occurs according to the given scenario, as a result of the interrelation of all elements of the preceding events with the initiating event (IE) in the structurally complex system [14]. Such scenarios are classified as functional failures of systems [15], which can be found in the RT by the "event trees" method [3, 15, 16] and—practically—using the FMEA, FMES, FTA, MEL standards and programs [17–19]. But in order to do this, the Prdf and pdf should be exactly specified in the RT, which is practically impossible with low statistics, and therefore, confidence limits are determined for the issues "of the rare events problem". Confidence-based Bayesian estimates do not allow the determination of exact "risk levels" if "risk is a probability". This was proven by NASA [3] and shown in Chap. 3 in paragraph 3.3.

4.2.2 Definition of Threats and Risks in SMSs

The provided recommendations on safety assessment and risk models are applicable for civil aviation: in justifying *operational safety requirements for aviation structures and systems* (in assessing the *airworthiness* of aircraft, helicopters, and spacecraft). The basis for identifying functional failures are diagrams of threats and risks and diagnostic models embedded in DBs of SMSs AA № 1, № 2.

It is assumed that the risk database is formed through chains of events based on the risk factor analysis diagram in Fig. 4.2 where phases 1 and 2 (identification and flight safety management) are shown.

These schemes are intended to be included in the SMS standards, taking into account the requirements of Annex 19 [1].

4.2.3 Use of Risk Analysis Matrices in Threats Analysis

The fuzzy measure $\mu 1$ of the random occurrence of a risk event R will be as follows: "rarely", "very rarely", "frequently", "infrequently", "sometimes", which allows the use of known risk assessment matrices by FAA, MES, etc. and avoid the incorrect concept of "predictable" ("guessed") probability of an event, which is accepted in many traditional SMSs—in the Russian Federation and in the USA [3, 6]. The integral assessment of the risk level $\hat{R}$ ("amount of hazard" defined as proposed above) as a function of a two-element set is found by (3.2)–(3.4) taking into account the amount of damage. The general principle of matrix construction and the adjusted ICAO matrix were given earlier in Chap. 2. Here, it is demonstrated by Fig. 4.3 that a row with an incorrect name for the concept "guessed probability" is excluded from the risk assessment matrix (e.g., in the SMM [7]). (In the JSC RR standards [20–22], this row still remains).

The line of fuzzy estimates of the possibility of a random occurrence of a risk event R (with the "near-zero" probability) for the Fuzzy Sets method is given in the form (4.5) (below) or in another, but similar form:

$$\mu_1 : (\text{"very rarely"}, \text{"rarely"}, \text{"infrequently"}, \text{"frequently"}). \qquad (4.7)$$

Phase 1. Identification of Threats and Risks

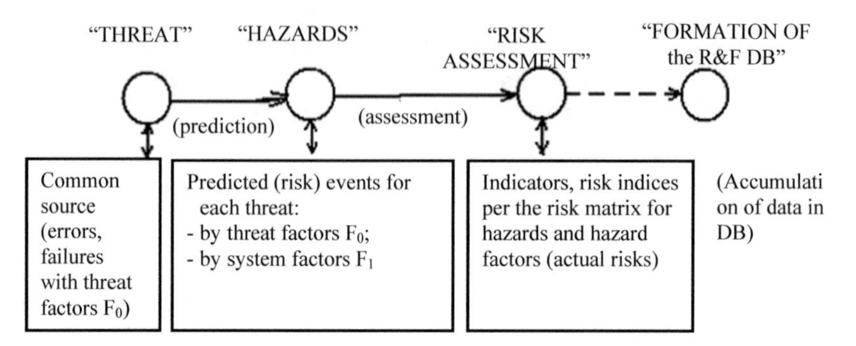

Phase 2. Flight Safety Management Through Risk Management

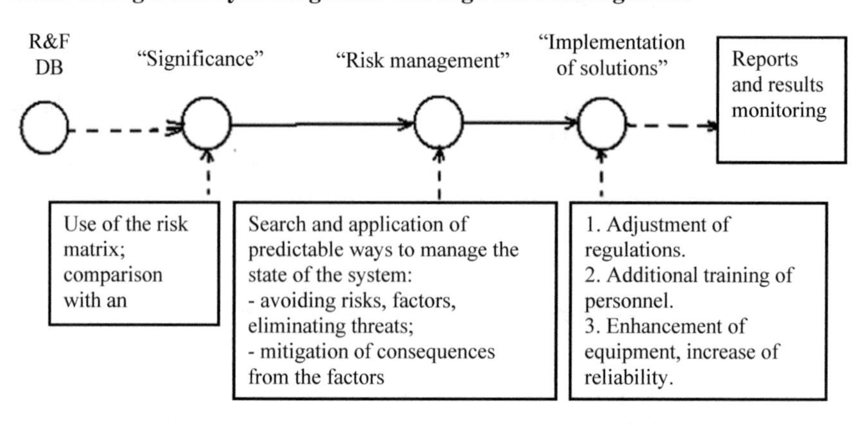

Fig. 4.2 Risk Factor Analysis Diagram

a) Initial matrix in the SMS
(A. Younossy – FAA, NASA)

"probability" in the SMS			
…….		…….	…….
……	**(α, β)**	……	……
……		……	……
……		……	……

b) Matrix adjusted in the RRS SST
(Fuzzy Sets)

μ_1 : ***"risk measure of the 1st kind" per (6)***			
…….		…….	…….
……	**(μ_1, β)**	……	……
……		……	……
……		……	……

Fig. 4.3 Transformed ICAO risk assessment matrices

The ICAO risk assessment matrices (in "SMS", "SMM", "NASA" documents) [3, 7, 23] were transformed and took the form shown in Fig. 4.3, which makes it possible to interpret their physical content.

In matrix cells, pairwise combinations of elements are arranged, as was originally suggested by NASA [3].

In this case, the concept of a scalar, of the *"medium risk"* type that is incorrect for rare events, is completely excluded from consideration. This is difficult to do in the PSA. However, in (4.5), (4.5), and (3.2)–(3.6) (from Chap. 3) are universal relations, where μ_I may be a PSA probability, if reliable statistics are available, and the "*average number of expected aircraft conflicts per hour*" in the ATC system.

4.2.4 Algorithm of the NASA Scenario for the Triad of Proactive and Predictive Safety Management for Aviation Activities by Means of SMSs (FO SMS–AA SMS)

In this subparagraph, the definitions included in the "triad" of Sect. 4.2.1 are explained.

On the basis of the risk concept from ICAO Annex 19, approaches to risk assessment and flight safety or service providers' activities safety management are proposed, including the industrial production of ATSs. This concept was adopted in this book, and methods for assessing risks ("amount of hazard") for specified scenarios of hazardous situations were developed.

NASA [3] proposed a universal (unified) algorithm for analyzing hazardous situations in the form of a tuple <…> of fuzzy attributes as follows:

$$< threat-risk\,event-prediction\,of\,events(leading\,to\,a\,catastrophe) \\ -hazardous\,state-risk\,assessment-control\,impact\; > \tag{4.8}$$

This algorithm is necessary for developing AA SMS (SMS) standards for risk and threat significance assessment—according to Fig. 4.2.

4.2.5 ICAO and ISO Hazard Models in SMSs

The general scheme for constructing hazard models is based on the methodology of Chap. 2 (paragraph 2.5) using the outcome space from the probability space (2.1) with the transition to the *Fuzzy Sets* methodology.

In the SST, the risk is defined as an "amount of hazard" in given critical discrete states with fuzzy subsets of analysis objects in some specially formalized structures of the systems under study.

Following this, it is assumed in the RRS that the primary system safety analysis should be performed using the concept of the system state in a discrete probability space for events with the "near-zero" probability. Then, according to the SST, the measurability of random events by probability is abandoned, and the fuzzy logic of the event algebra is adopted using the *Fuzzy Sets* method. The prediction (proactive and predictive) methods of safety regulation by ICAO are implemented on the methods of calculating the risks of occurrence of negative consequences in technical systems that are detected by SMS-like systems.

The main task in the general system safety theory is to define and interpret the category of "residual risk" arising from the RT concept and to recalculate this "risk" with the help of SST tools into residual "catastrophe risks" that determine "passive" and operational safety in highly reliable systems.

ICAO hazard models can be found as event chains, clear in functions, in the form of scenarios. Then, according to the SST method, it is recommended to *find, without probabilistic indicators, the risks* of negative consequences *from these scenarios*. To this end, Annex 19 recommended to use risk assessment matrices in (Fig. 4.3, in general terms).

4.3 Construction of a Generalized Safety Management System (SMS)

The main features of constructing an international SMS are most clearly stated in the document created in the US Federal Aviation Administration (FAA) service under the leadership of Amer Younossi, head of the FAA Flight Safety department. This document ("Blue Folder") presents the main modules of the system recommended by ICAO for inclusion into the new international standards such as Annex 19. The "10 positions" (from "Blue Folder" [4]) are indicated, which should be reflected on the basis of this document (shown in Fig. 4.1).

4.3.1 SMS Functions Based on the NASA Principles (for ICAO)

In the generalized SMS, SMSs should be presented in the form of two modules: AA SMS—Type 1 and Type 2.

In Type 1 AA SMS (state/national level), the following tasks are solved:

- Continuous monitoring of the aircraft (all aircrafts in the region);
- Real-time optimization of the flight plan, the use of power plants, taking into account the current meteorological, aerodynamic and statistical data for a specific aircraft;

- Interaction with ground technical services is provided during the flight; thus, reducing the time for maintenance of the aircraft after the flight is completed and for its preparation for the next flight;
- Aircraft systems and units operation are monitored with remote diagnostics provided;
- An automated system for collecting and analyzing the information received is used;
- Fail-safe communication channel on 100% of the surface of the globe;
- Immediate notification of the crew and ground services of failures and deviations in the operation of the onboard systems;
- Analysis, notification, and recommendations on fuel efficiency at any time, at any flight phase;
- Conclusions based on the analysis of accumulated information;
- Database replenishment and administration;
- Current actual level of aviation equipment reliability;
- Automated detection of factors of possible hazard occurrence along the flight route and advance notification of the aircraft crew;
- Optimization of aircraft maintenance;
- Integration of transmitted data;
- Continuous diagnosis of possible aviation equipment failures;
- Timely messages from the crew about the need for unscheduled works to eliminate defects detected in the course of the flight;
- Crew access to ground information systems and resources;
- Consultations with ground services during the flight;
- Transmission and exchange of data between aircrafts; and
- Receipt of information on special flight conditions from the aircraft.

To ensure the effectiveness of these actions, databases are compiled containing common threats, particular threats by hazard factors, risks, a list of risk management functions (chains); the risk values are adjusted; and the impact on the ATS is estimated by means of the SMS in airlines.

4.3.2 Principle of Constructing and Determining the Composition of the AA SMS (Type 2) Core

The main unit of the AA SMS is the "core" in (4.1) consisting of separate modules that make it possible to evaluate and improve the overall level of safety: by identifying characteristics of adverse single and rare events with small statistical samples and to take measures to eliminate and prevent them.

As the main modules of the "core" providing proactive safety management schemes (shown in Fig. 4.2) and the output results from service providers, the following are taken, according to (4.8):

- *identification of threats (list of hazards, events, and factors);*

- *identification of risk events (risk assessment);*
- *determining the consequences (assessment of damage from the knowledge base); and*
- *identification of management actions that effect the system to minimize risks.*

These modules are, in essence, the same as in the SMMs of early editions [2]. But now their number is minimized and stipulated in the new ICAO standard (Annex 19).

The concept of risk in the presented system is interpreted in the original form as it was indicated in Chap. 2 and adopted in the engineering and physical sciences as follows: "*Risk is a possible hazard*". This hazard is caused by threats: management errors, human factors, etc. In the methodology for calculating risks, hazard is always considered *as predictable* if the conditions for the occurrence of a risk event are detected. The *problem is the detection of hidden threats in the system*, depending on the residual risk designed at the stages of development and production of transport equipment in general and aviation equipment (aircraft, helicopters) in particular.

4.3.3 SMS Subsystems and Modules

According to ICAO (Annex 19) [2], SMS should be built as a general system of 2 subsystems global for the country (region). Accordingly, the general system, AA SMS, consists of two subsystems:

SMS-1—State/national level (or sectoral);

SMB-2—Service provider level (airline, airport, maintenance and repair, "Production").

The first subsystem SMS-1, operating at the state level of aviation activities (AA), performs the following functions:

- continuous monitoring of flight parameters and system states of all aircraft in the region (this has always been done in civil aviation of the Russian Federation, but on a different technical information basis and for other purposes);
- creation of an interaction "aircraft—ground technical services" system on the basis of the ACARS prototype (A-380) to be used during the flight and during aircraft maintenance; and
- replenishment of databases and administration of all service providers.

The second subsystem SMS-2, which is part of the general SMS structure, contains two main standard modules:

Module 1—Integration of QMS units with modules that determine the principles of SMS functioning, ensuring the achievement of the main result—providing a target level of safety based on the concept of risks (per ICAO), taking into account all aspects of the provider's activities.

Module 2—Tools for measuring and predicting risks, assessing the integral significance of risks, normalizing acceptable risks, formalizing and creating a system for identifying risks and a base of risk factors, managing risks according to scheme (4.8), taking into account (4.7), and preventing occurrence of negative situations in accordance with the recommendations of regulatory documents.

The global AA SMS (and database, DB) created on the basis of SMS-1 and SMS-2 on the NASA prototype, as required by ICAO, performs the following functions:

- Automated collection and analysis of information received onboard the aircraft;
- Analysis of aircraft communications on the entire surface of the globe (FORAS, ACARS and in ground services);
- Processing of the accumulated information from the perspective of analyzing risks and quality of operation of systems and units on the basis of quality standards (reliability of units, hazardous scenarios, etc.);
- Evaluation of the current level of reliability and safety of aviation equipment during the life cycle on the basis of relevant manuals;
- Automated detection of hazard factors along the flight route and advance notification of the aircraft crew of possible threats, activities of the service provider, resources, vehicles, etc.

Note Classification of threats, list of hazards, and hazard models, for example, in the form of scenarios and J. Reason chains, should be defined in another separate standard.

The generalized functional diagram of the SMS-2 "core" is shown in 4.3a. A detailed description of this diagram was given in Fig. 4.3. [5, 10] (above).

4.3.4 Functional SMS Diagram and Computer Support of Procedures for Assessing Risks of Occurrence of Adverse Events on the Basis of the ICAO Methodology (SMM)

The SMS prototypes and their characteristics are described by airline "British Airways" on the example of Bristol airport; USA FAA—"Order", as well as in the IATA "Hand Book" and other documents. Well-known service providers (FAA, Canada, Aeroflot, Transaero, Russia, etc.) also have prototypes of such SMSs.

Here, for the first time, the Fig. 4.3 is showing a generalized SMS diagram, which should be included in a series of standards, developed for presentation to ICAO and for civil aviation in Russia so as the form of national standard.

The main SMS modules can be as follows.

- The module of the SMS organizational structure and its subsections should be created in accordance with the ICAO SMM recommendations, primarily on the basis of Annex 19 [1, 6].
- Mandatory proactive flight safety management modules based on risk factors recommended by the SMM [24] and defining the functional composition of the SMS.

- Modules (per ICAO) that implement mandatory procedures for identifying risk factors and a list of hazards and threats [7, 24, 25].
- Information databases (IDB) for proactive risk identification based on statistics (FI decoding) and predicted threats.
- Bases of expert knowledge, which are grouped into the "portfolios" of risk analysis matrices with hypertexts of the predicted consequences.
- Algorithms for developing proactive corrective actions that affect the state of the system.
- Procedures for identifying and classifying simple risks, J. Reason chains, detecting hidden threats (in the form of filters in the protection equipment).
- SMS system of computer support for results of the flight safety monitoring with the formation of integrated flight safety indicators for service providers as follows:

– integral estimation of the flight safety state in real time with the scientific and methodical modules predicting hazards;
– feasibility study of the decisions on factor risk management in the service provision system (airfields, ATC, fuel, etc.).

- A system for displaying information in a generalized form for providing information to the management of the service provider organization and the civil aviation authority of the Russian Federation in the ON-LINE mode and for monitoring the flight safety state in civil aviation of the Russian Federation from the part of ICAO.

In Russian Civil Aviation, there are some examples of SMS, created on basic SMM principles.

4.4 Methodological Basis for Solving the Problem of Estimating Residual Risk Taking into Account ILS Chains

4.4.1 State Regulation of AA Safety in Civil Aviation of Russia

The state prioritizes the level of acceptable risk $\hat{R}_{*i}$ for occurrence of emergency situations in civil aviation, according to the flight safety program (Order No. 641r of 06.05.2008).

It is assumed here that the concepts of "risks" and their significance mentioned in various documents are set in terms of the RRS included in the SST tools, i.e., as the universal concept of "integral risk" $\hat{R}$ from Chap. 2. This, in particular, is the convenience of the approach to solving safety issues on the basis of the RRS provisions [26–28].

4.4.2 Determination of Acceptable Risk Levels

The concept of "acceptable risk" was originally introduced in the RT—in I. Aronov's book [16, 29] and in other publications (this book contains this concept cited from [16] in Chap. 1). This term is recommended to be applied both in the state regulation of flight safety and in all aviation activities, including the ATS production (since 2010, also in JSC RR [29]).

Taking into account the results achieved in previous $(i-1)$ periods of operation and production of aviation equipment, an acceptable risk is assigned as:

$$\hat{\mathrm{R}}_{*\mathrm{i}} = \frac{1}{k}\,\hat{\mathrm{R}}_{*\mathrm{i}-1}, \tag{4.9}$$

where $\frac{1}{k}$ is a coefficient of hazard reduction in the adopted measurements in risk units due to the corresponding increase in the level of flight safety at the ith stage of the civil aviation industry cycle. In particular, for the period up to 2012 the coefficient (1/2) was introduced, i.e., the acceptable risk should be reduced 2 times:

$$\hat{\mathrm{R}}_{*\mathrm{i}} = \frac{1}{2}\,\hat{\mathrm{R}}_{*\mathrm{i}-1}. \tag{4.10}$$

For example, according to ICAO, in Russia the number of accidents or emergency situations should be reduced by half. This implies the requirement to improve the quality of aviation equipment production and its operation, taking into account the ILS principles. This requirement leads to the need for redistribution and recalculation of risks across the entire ILS production chain, so that in the end, as a whole, an acceptable level of flight safety is ensured in civil aviation of Russia, as established by the state through the level of risk.

The question is, *if the PSA method is used, then which units this "risk" or flight safety level is to be measured in?* In the RRS (SST), this issue is solved with the help of $\hat{\mathrm{R}}_0$.

The "residual predicted risk" is determined by production through a fuzzy measure of the predicted hazard, for example, by applying quality systems that clearly give confidence bounds for the probability of occurrence of functional failure states in a range of the failure event probability up to 10^{-6}–10^{-3}, not worse.

Integral indicators of an acceptable level of risk $\hat{\mathrm{R}}_{*i}$ established by the state, for example, ICAO, in the form of 4–5 or 1–2 catastrophes over 10–15 years of the aircraft operation should be recalculated into residual risk indicators R_{M} of the equipment transferred to the operator.

However, the problem is that equipment manufacturers ensure the quality of their ATS according to the standard probability indicator only. The manufacturer assumes by default that "*if the AE is reliable, then it is safe*". But this condition of flight safety does not meet the requirements of international standards. In Chaps. 2 and 3 of the book, it was shown that "catastrophes" (in the RT) with *"non-zero" residual design risk* $\Delta\hat{\mathrm{R}} \neq \Delta\hat{\mathrm{R}}_*$ are not eliminated, but only moved to "infinity" according to the rules of "3" or "6". This is not constructive for civil aircraft, NPPs and HPPs because of small production volumes only the production of high-reliability cars (e.g., TOYOTA) has indicators $\Delta\hat{R}_*$ close to the probability of a possible risk event R_*, the occurrence of which is caused by the rule of "6". Here, (3) and (6) rules are indicated: *"Three sigmas" and "Six sigmas"*.

At present, manufacturers have not any methods for converting risk factors into reliability indicators, nor are they obliged to modify their production processes in any way.

Thus, the *"Manufacturer-Operator" system* is open (in Russia) in terms of flight safety indicators. While Western systems have been adhering to the well-known ILS principle and supplying equipment to Russia for a long time following the after-sales service method. At the same time, the ILS principle is observed quite fully in accordance with the adopted regulations.

It is assumed that operators' risks $\hat{R}_{\mathrm{O}}$ are functions of the parameters n_i of the flight operations system in airlines,

$$\widehat{R}_{\mathrm{O}} = f_{\mathrm{O}}(n_1, n_2, n_3, \ldots n_i, \ldots, n_{\mathrm{O}}) \tag{4.11}$$

Manufacturing risks $\hat{R}_M$ are determined by a list of interrelated parameters k_i characterizing the ILS system (ILS parameters), for example, per IATA:

$$\widehat{R}_{\mathrm{M}} = f_{\mathrm{M}}(k_1, k_2, k_3, \ldots, k_i, \ldots, k_{\mathrm{O}}) \tag{4.12}$$

The difference (discrepancy) $\Delta\hat{R}_{\mathrm{OM}}$ of risks $\hat{R}_{\mathrm{O}}$ and $\hat{R}_{\mathrm{M}}$ should not exceed the minimum level $R_{\min}$ determined by an acceptable level $R_{*\mathrm{i}}$ as established by the state:

$$\hat{R}_{\mathrm{M}} - \hat{R}_{\mathrm{O}} \leq \hat{R}_{*i} = \frac{1}{2}\hat{R}_{*i-1}, \tag{4.13}$$

which is equivalent to the condition indicated above,

$$\Delta\hat{R}_{\mathrm{OM}} = \hat{R}_{\mathrm{M}} - \hat{R}_{\mathrm{O}} \leq \hat{R}_{*i} = \frac{1}{2}\hat{R}_{*i-1}, \tag{4.14}$$

When using risks in the format of indicators (ICAO, IATA), these ratios determine the principle of proactive management of the system, taking into account the ILS and flight safety indicators in the aircraft operation, as shown above:

$$\Delta I_{\hat{R}12} = I_{\hat{R}2} - I_{\hat{R}1}, \quad \Delta S_{\hat{R}12} = S_{\hat{R}2} - S_{\hat{R}1},$$

where $\Delta I_{\hat{R}12}$ is the discrepancy between the risk indicators, and $\Delta S_{\hat{R}12}$ are additional costs to cover risks in airlines and in the aviation industry, necessary to achieve an acceptable risk level in the form of (4.9).

Thus, it is necessary to substantiate the selection of hazard factors in the two subsystems (“manufacturing”, “operation”) and determine the principles of safety regulation.

In addition, a set of documents of the international level should be established to implement the methodological foundations for the construction of a state system to ensure flight safety in the Russian Federation.

4.5 Conclusions

1. The ICAO recommendations (in Annex 19) stated the need for mandatory development of SMSs for various providers’ aviation activities. At the same time, there are some unified requirements for the functions and the minimum composition of SMS modules. The NASA algorithm for providing proactive risk management in ATSs is universal.
2. In order to eliminate contradictions in the interpretation of the functions, composition, and purpose of the SMS, it seems advisable to develop a national standard for civil aviation of the Russian Federation regarding the concept of constructing a unified SMS (AA SMS), in which the “core” and specialized databases should be defined that reflect the specifics of various providers’ aviation activities.

References

1. Annex 19: C-WP/ 13935—ANC Report (March 2013), based on AN-WP/ 8680 (Find) Review of the Air Navigation Commission, Montreal, Canada
2. SMM: Doc. 9859-A/N460. ICAO, 2009
3. Probabilistic Risk Assessment Procedures for NASA Managers and Practitioners—Office of Safety and Mission Assurance NASA. Washington, DC 20546—August. 2002 (Version 2/2)
4. Amer MY (2012) 10 Things you should know about safety management systems (SMS). SM ICG, Washington
5. Evdokimov VG (2012) Integrated safety management system for aviation activities based on ICAO standards and recommended practices, J TR, 2(45):54–57. (in Russian)
6. SMS & B-RSA (2008): “Boeing”, 2012
7. SMM (Safety Management Manual): Doc 9859_AN474—Doc FAA: 2012
8. Risk management (2009)—Vocabulary—Guidelines for use in standarts. PD ISO/ IEC, Guide 73: 2002.B51
9. Gipich G, Evdokimov V, Kuklev E, Mirzayanov F (2013) Tools: identification of hazard & assessment of risk. report at the ICAO meeting “Face to Face”. “Boeing” Corp., Washington, janv
10. Kuklev EA, Evdokimov VG (1995) Predicting the safety level for aviation systems based on risk models. Journal TR, No. 2 (45), pp. 51–53. (in Russian)
11. Kabanov SA (1997) System management on predictive models. St. Petersburg University, St. Petersburg (in Russian)
12. McCarthy J (1999) U.S. Naval Research Labaratory; Schwartz N, AT & T. Modeling risk with the flight operations risk assessment system (FORAS).—ICAO Conference in Rio de Janeiro, Brasil, Nov

13. Documents of the 37th ICAO Assembly (October 2011, Montreal)
14. Volodin VV (ed) (1993) Reliability in technology. Scientific-technical, economic and legal aspects of reliability.—Blagonravov Mechanical Engineering Research Institute, ISTC "Reliability of Machines"—RAS, Moscow, pp 119–123. (in Russian)
15. Aronov IZ et al (2009) Reliability and safety of technical systems. Moscow (in Russian)
16. Smurov MY, Kuklev EA, Evdokimov VG, Gipich GN (2012) Safety of civil aircraft flights taking into account the risks of negative events. J Trans Russ Fed, 1(38):54–58 St. Petersburg: 2012. (in Russian)
17. Smurov MY, Kuklev EA, Evdokimov VG, Gipich GN (2012) Development of tools for assessing the risks of occurrence of risks of AUI in the AASS of the airport system. J Trans Russ Fed, 2(39), St. Petersburg (in Russian)
18. Evdokimov VG, Kuklev EA, Shapkin VS (2013) Use of flight information to increase the reliability in assessing the levels of possible threats. Scientific bulletin of the MSTU CA. No. 187 (1). Moscow, pp 49–53. (in Russian)
19. Safety provisions (2010) Helicopter European safety group and JSC ALLIED AVIATION CONSULTING—EHEST (ESSI). JSC "AAC", Moscow, 2010. (in Russian)
20. Guidance on hazard identification. SMS WG (ECAST (ESSI) Working Group on Safety Management System and Safety Culture). Handwritten, 2010. (in Russian)
21. General rules for risk assessment and management (resource management at life cycle stages, risk and reliability analysis management—URRAN). JSC Russian Railways standard STO RR 1.02.034-2010. Moscow, 2010 (in Russian)
22. Documents of the ICAO High-level Conference (SMM). March 2010, Montreal
23. Issues of the system safety and stability theory./Issue 7. Ed. by Severtsev N.A./Moscow: RAS Dorodnitsin CC, 2005. (in Russian)
24. Accident Prevention Manual (1984) Doc. 9422-AN/923. International Civil Aviation Organization
25. Evdokimov VG (2010) Modern approaches to safety management based on the risk theory. International Aviation and Space Magazine "AviaSouz". NQ 5/6 (33) October–December 2010. (in Russian)
26. Evdokimov VG, Gipich GN (2006) Some issues in the methodology of assessment and prediction of air transport risks. In: The 5th international conference "Aviation and Astronautic Science 2006", Abstracts, Moscow: 2006. pp 75–76. (in Russian)
27. Evdokimov VG, Grushchanskiy VA, Kavtaradze EA (2008) On the system safety of information support for the development of complex social systems. Fundamental problems of system safety: Coll. of papers/RAS Dorodnitsyn CC, Moscow: University Book, pp 67–77. 14. (in Russian)
28. Aronov IZ (1998) Modern problems of safety of technical systems and risk analysis. Standards and quality, vol. 3. (in Russian)
29. Procedure for determining the acceptable level of risk (URRAN). JSC Russian Railways standard STO RR 1.02.035–2010. Moscow: 2010. (in Russian)

Chapter 5
Algorithms and Methods for ATS Safety Monitoring and Assurance Using Methods for Calculating Risks in the RRS Doctrine

5.1 Methodical Provisions for Assessing Aircraft Operation Safety

5.1.1 Definitions of Risk Varieties

The following main provisions are adopted in the formulated problem.

- The concept of flight safety assurance is adopted on the basis of the definition of the term "Safety" according to ISO-8402, Annex 19, and the corresponding mathematical risk models ("hazard measure", "amount of hazard", "hazard", but not "probability".).
- The methodological basis of computational technologies and procedures for risk assessment and selecting an acceptable risk is the use of certain tools: "Chains, events in risk situations, risk events", "risk evaluation", "risk management, management objective, achievement of acceptable risk", "prevention of catastrophes and accidents taking into account the estimated (potential) risk", "accident avoidance on the basis of risk assessments and risk management", predicting of a "hazard measure"—according to the NASA scenario in the form of "tuple" (4.8) from Chap. 4.
- Methodological provisions on hazards and occurrence of risk situations in the aviation system during flight operations and in assessing aviation activity safety are based on a threat (cause or IEs) prediction diagram and the expected consequences:

  (a) Absence of a threat (source of hazard) denotes the existence of a clear "state of safety" (in the sense that there is no threat or it is not declared or a threat is not detected, that is, the state of hazard does not arise, i.e., the magnitude of risk is not estimated: No threat—No risk of negative consequences in the system);

Kuklev E.A. et al., *Aviation System Risks and Safety*, Springer Aerospace Technology, https://doi.org/10.1007/978-981-13-8122-5_5

(b) For each type of threat (type 1a or type 1b), the predicted outcomes form the minimal binary space of predicted outcomes: risk event (“risk”) and (“chance”) [1–6].

5.1.2 Characteristics of Hazardous States of Systems

The characteristics are as follows.

(a) A threat is a source of hazard with certain factors affecting the aircraft and the state of the flight situation.
(b) A hazard is a state of the system that occurs when hazard factors manifest that can lead to negative (undesirable) consequences.
(c) The safety level is determined by comparing the values of possible (predicted) risk of undesirable consequences with the level of acceptable risk for a specific type of hazard (and factor) for a given (identified) threat.
(d) The significance of the calculated risk level is determined expertly by means of risk analysis (assessment) matrices in the form recommended by ICAO and presented in the SMM and adopted as a basis in the SMS-R for airlines [7].

5.1.3 Methodical Provisions of “Preventive” (Proactive) Hazard Prediction in Order to Improve Flight Safety Based on Risk Management Through ATS Parameters Taking into Account Risk Factors

The risk or magnitude of the risk in the sense of the adopted definitions in terms of the “amount of hazard” (or hazard significance) is formulated as described above in the SST.

(a) The significance of the risk of undesirable consequences of the predicted flight results is estimated on the basis of the proactive method using a risk matrix in the form of a single definite number or an index reflecting the expert value of a combination of numbers providing, in a fuzzy measurement of the frequency of occurrence and the severity of consequences or damages, the idea of predicted results for given factors and identified threats.
(b) The acceptability of the calculated proactive risk is found by comparing it with the allowable (standard) level of acceptable risk so that the discrepancy of the two levels—calculated or predicted and allowable or acceptable—is used to develop management actions on the system in order to reduce the calculated predicted risk and to improve the safety level in airlines.
(c) Risk management and safety management are provided through effects on the system by factors and conditions of occurrence of hazardous events, with spe-

cific hazard types and factors and with identified (proactively detected) threats (sources of hazard).

(d) Risk mitigation methods based on the management of the aviation system state include avoiding hazard factors (sources of hazard) or elimination of sources of hazard.

5.1.4 Methodological Provisions on the Relationship Between the Characteristics of Proactive and Active Methods for Assessing the Significance of Hazards and Risks for the Factors Database and the List of Hazards of a Particular Airline

(a) A proactive method of safety management based on risk management is the key to increasing the level of ATS safety by predicting in advance—a priori (proactively)—of possible negative consequences from each identified (proactively or actively) hazard factor and the corresponding source of hazard (threat).

(b) An active method for risk and flight safety management is used in the investigation of accidents to detect hidden or unknown and undetected threats (sources of hazard) and to confirm and verify expert assumptions made in proactive risk identification methods.

(c) Operations to manage flight safety by risk management are a chain of sequential actions in the NASA tuple (4.8) from Chap. 4.

5.2 Tools for Identifying and Estimating Risks in Solving the Rare Events Problem Within the New Doctrine "Reliability, Risks, Safety"

5.2.1 SST Tools. *The List of Tools Includes the Following*

- procedures, algorithms, methodologies, guidelines following from the fundamental positions of the new doctrine;
- standards harmonized with international requirements and the provisions of the new RRS doctrine;
- guidelines and methods for proactive (per ICAO) management of the system safety state based on the methodology of calculating the significance of risks, taking into account surveillance data and prediction of sources of threats and hazard factors in Fuzzy Sets of signs of the significance of hazards;

- a typical functional diagram of the SMS established by the standard on the risk management for rare hazardous (risk) events of "functional failures" with given scenarios.

Proactive management means that the system is influenced through its main parameters in advance (for the predicted threat) until the moment this predicted possible risk event occurs. The methodology of calculating risks in the "rare events" problem has determined the current trends in the SMS improvement [8–10].

5.2.2 *Basic Principles of Flight Safety Management*

- Flight safety management is provided with the use of ICAO and IATA standards;
- The current state of flight safety is controlled and monitored on the basis of monitoring the indicators of continuing airworthiness during the aircraft operation, maintenance, and repair on the basis of diagnostics of its technical condition and the use of ACARS systems (as for A-320, A-380 aircraft);
- The minimum sets of parameters that determine the significance of risks $\hat{R}_O$, $\hat{R}_M$ (and for "residuals" (discrepancies) of risks ΔI) are assigned per ICAO Annexes and per IATA from the IOSA NBP;
- The minimum number of various standards necessary for the implementation of this methodology is determined by the state aviation administration;
- The IOSA document set is adopted as the normative baseline, in which there are currently 62 standards, including those on the problems of servicing for aircraft such as "Airbus" and "Boeing" in the Russian Federation;
- The principles of calculating risks in the "rare events" problem adapt to the recommendations of the new doctrine "Reliability, Risks, Safety" for recalculating indicators of the acceptable risk R_{*i} established by the state into *residual risks* of emergency situations in aviation equipment produced in the aviation industry on the basis of the *standard reliability indicators*.

5.2.3 *Concept of Constructing J. Reason Chains in Fuzzy Subsets of ATS States*

Algebra of the clear events logic in the scenarios of catastrophes determines the nature of these catastrophes. The ACARS monitoring system in civil aviation offers trigger simulation modeling and combinatorial analysis of events for a given fuzzy set of threats of various levels for procedures of predicting possible consequences in J. Reason chains. Algorithms for predicting the logical conditions of catastrophes are based on the use of the principle of minimal cut sets and functions of the operable system. Hazard and risk models per ICAO (NASA) are developed on the basis of methods for analyzing combinatorics of events in sigma-algebra in probability spaces with an estimate of fuzzy consequences. Algorithms and recommendations on the methodology for assessing the degree of risk in comparison with the level of acceptable risk are applied.

5.3 Determining and Assessing the Significance of Risk for Events from the Space of Binary Outcomes Using Risk Analysis Matrices

5.3.1 Types of Risk Matrices Per ICAO

The known concept of risk of ICAO [7, 11, 12] contains definitions of risk concepts and its significance for ATS and AA that allow the use of the RRS doctrine and SST tools to determine flight safety indicators through hazard levels in the form of integral risks $\hat{P}$ that depend on the hazard model in the form of $\tilde{P}$, as in (3.2). The assessment of risk $\tilde{P}$ as an amount of hazard in a system with a predictable risk event P is initially specified either by a set of indicators [8, 12] or in integral form $\hat{P}$, for example, in points or indicators, or using risk analysis matrices [8, 13] shown in Fig. 5.1, taking into account the prototypes in Fig. 5.2a–c.

NASA MATRIX for RISK ANALYSIS

Likely - hood of risk event	Severity of risk				
	Catastrophic A	Dangerous B	Significantly C	Insignificantly D	Pitiful E
Often 5	5A	5B	5C	5D	5E
Sometimes 4	4A	4B	4C	4D	4E
Very rarely 3	3A	3B	3C	3D	3E
Improbable 2	2A	2B	2C	2D	2E
Extremely rare 1 improbable	1A	1B	1C	1D	1E

Fig. 5.1 Adjusted ICAO matrix

(a)

Row: Degree of protection (*Aviation safety level assurance*) – by classes A, B, C, D, E – *elements* $\{a_i\}$.
Column: Possibility of AUIs in Aeroflot activities (***Threat Level***) – by categories 1, 2, 3, 4, 5 – *elements* $\{b_j\}$.
Risk indices (AUI) are table cells $c_{ij} = (a_i, b_j)$

The hazard of terrorism manifestations

(b)

"Boeing" (IAC, 2003)
CD-Guide (51G-BASP Safety Inipl.ppt)
"Risk is the combination of the severity of a specific hazard and the likelihood that the hazard will occur"

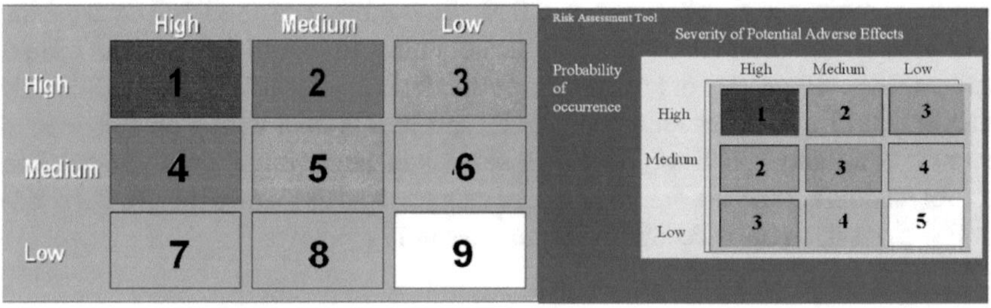

	High	Medium	Low
High	1	2	3
Medium	4	5	6
Low	7	8	9

Risk Assessment Tool

Probability of occurrence	Severity of Potential Adverse Effects: High	Medium	Low
High	1	2	3
Medium	2	3	4
Low	3	4	5

Fig. 5.2 **a** Matrix of risks of AUI occurrence—"Boeing". **b** Risk models for two-dimensional estimates (Boeing, Aeroflot)

5.3.2 *Binary Partitions of the Outcome Space in the Risk Analysis Matrix*

The mathematical characteristic reflecting the physical essence of risk follows from the concept and binary space Ω of outcomes ω_0, ω_1:

$$\Omega = \omega_0 \cup \omega_1 \cup \varnothing, \quad \omega_0 = A, \quad \omega_1 = \bar{A} \equiv R, \tag{5.1}$$

where ω_0 is a class (set) of events that are not hazardous, $\bar{A}$ is the event opposite to ω_0, i.e., hazardous, for example, risk event R such that $R \equiv R_\Sigma = \cup R_i$ is a class of events in the group R_Σ [1, 14].

Relation (5.1) determines the practical meaning of the ICAO risk matrices (according to "Boeing") from [8]. The traditionally recommended matrix provides the value of randomness and damage for one event only—for a critical outcome in the form of a risk event $\bar{A} \equiv R \approx R_\Sigma$.

Since, as known, risk (hazardous) events are rare, we cannot obtain a detailed parent entity and thus be limited by the recommendation (5.1). In view of the rarity of the events from the class ω_1 in (5.1), it is necessary to evaluate expertly the measure of randomness of this event ("guess" it, as in [15] for tankers). In this case, there will be one matrix, and several results of assessing the significance of risks in points, for example.

Formulas for risk assessment were proposed in Chaps. 1, 2—per the PSA and the RRS.

The magnitude of the risk as a physical category or its estimate $\tilde{R}$ are formally, according to the ICAO concept, estimated through a two- or three-dimensional set of indicators [9, 12] in the form of (3.2), (3.3) from Chap. 3. Copies of these results are given below:

$$\tilde{R} = \left\{\mu_1, \tilde{H}_R|\Sigma_0\right\}, \tilde{R} = \left\{\mu_1, \mu_2\tilde{H}_R|\Sigma_0\right\}, \tag{5.2}$$

where μ_1 is a measure of risk of the first kind (the uncertainty of occurrence of a negative result) in the form of a randomness or uncertainty indicator of occurrence of a risk event with a predictable (clear or fuzzy) measure of the possibility of its occurrence. This value can be measured expertly without a probabilistic category, in contrast to [15], which is more appropriate; $\tilde{H}_R$ is a measure of the consequences or damage (cost or amount of risk); μ_2 is a measure of risk of the second kind in the system due to system errors; Σ_0 denotes the conditions of the experience or situation in the operation of the system (hazard class or model of the system), the scenario of the development of events in case of an accident or catastrophe, and this condition (and sign of Σ_0) is mandatory, as it contains assumed hypotheses about models of threats and adverse factors in systems with possible manifestation of "*signs of predictable or active threats*".

Here, μ_1 is practically an expert indicator with some measure of randomness of events (*rarely, frequently,* etc.) as frequency from statistics of events or an indicator from risk matrices as in Fig. 5.1.

Alternative methods of risk management (according to Annex 19):

- *full compensation of all estimated losses and damages*—without risk management (*insurance, reinsurance of risks*);
- *"non-acceptance" of risks* (non-acceptance of negative consequences from possible risk events) by "*avoiding risk factors*" (i.e., the consequences of possible "manifestation of adverse factors" in detected or predicted threats);
- strategy of *partial acceptance of "calculated risks"*, etc.

The a priori *categorization of critical transport infrastructure (civil aviation) facilities* by analogy with the RT "event tree" method and "functional failures" is presented by "Boeing" in Fig. 5.2a, b, but initially the concept of "probability" was adopted in the matrix in Fig. 5.2, which is incorrect.

In this scheme, it is assumed that the risk of possible violation of operating modes in the state of ATS "functional failures" (according to M. Neymark) [16] is a single criterion for taking measures to protect AE at all levels of the ATS safety management hierarchy.

An airport of the first category, for example, has such critical elements as a terminal building, aircraft, an aircraft parking stand, an air traffic control unit. An ATS of type "helicopter" has critical elements such as "swashplate", "propeller blades", "gearbox", "tail rotor", etc. In the design of the aircraft and in the maintenance and repair infrastructure, as well as in aircraft servicing, there are critical elements failures of which are a possible cause of catastrophes. This fact is established a priori at the stage of identification of threats and possible risks in systems.

The required protection profile of each critical element is established by the criterion of the adequacy of protection with an acceptable level of risk of violation of the normal state of critical elements, for example, through the cost of the proposed protection system, taking into account the degree of protection in the risk matrix from Fig. 5.1.

5.4 Methodology for Assessing the Degree of Risk in Comparison with the Level of Acceptable Risk

5.4.1 Initial Provisions of the Adopted Methodical Approach

The need for an analysis of the formulated issue is conditioned by the criticism (from the positions of Annex 19) of the "rigorous" definition of the concept of "safety" in GOST R ("Safety aspects" [17] of 2002 as: "*Safety is a state of the absence of unacceptable risk*". The world community abandoned this formulation, but in Russia (even in aviation universities) this ill-conceived concept still exists.

The provisions adopted in the methodology of this issue are as follows:

- Any deviation of the calculated (potential) risk α from the acceptable level α^* is considered unacceptable regardless of the magnitude $\Delta\alpha$ of the exceedance of α from the standard value α^*;
- The definition of flight safety and the concept of risk developed in the book on the basis of Annex 19 are considered equal in terms of legal norms and are based on the assessment of the significance of relevant indicators in standard calculation procedures;
- An acceptable level of risk for ATSs introduced by ICAO in the concept of "Safety … through the state and risk" is assigned taking into account risk classes according to the following:

$\alpha^* \sim \alpha_1^*$ or α_2^*, …. etc.

It turns out, for example, that in relation to the flight safety level:

at $\alpha < \alpha^*$ *flight safety is provided*,

at $\alpha \geq \alpha^*$ *there is an unacceptable risk, the flight safety level is unsatisfactory (protection is necessary)*, i.e., the use of system protection means against the manifestation of threat factors, for example, based on the SMS.

It should be noted that in the *theory of law* there is no concept of risk, but there are symbols of "*threat*" and "*hazard*", therefore, international standards and ICAO SARPs are applied in civil aviation of the Russian Federation.

5.4.2 Graded Classes of Fuzzy Risk Boundaries ("Granules")

The concept of "granules" is introduced here from the publications of I. Grozovskiy from the All-Union Research Institute (GOST-R) [18]. The flight safety levels by risk classes are as follows into (5.3):

$$\left\{\begin{array}{l} \alpha < \alpha_0^* \text{—``accepted''—} \textit{minimum, but allowable risk;} \\ \alpha < \alpha_1^* \text{—(class 1)—} \textit{allowable} \text{ in a particular situation;} \\ \alpha < \alpha_2^* \text{—(class 2)—``} \textit{tolerable risk} \text{``;} \end{array}\right\}$$
$$\alpha < \alpha^{**} \text{—} \textit{unacceptable risk}, \text{etc.} \tag{5.3}$$

With this approach, the *flight safety is defined through a state* in which *there is a certain set of "protection means"*, programs and methods for the *appropriate class of risks*. The definition of flight safety adopted through *"something" that "exist" is more logical and constructive than through the "absence of something"*, when measures are to be taken or a device is to be taken out of use, such as a "*«Toyota» car with faulty brakes*".

Thus, *with this approach, it is important that there are (obviously) some means to ensure flight safety with* α_i^*, *and something that "does not exist" is of no importance.*

Consequently, the wording from Chap. 2 can be accepted that "*A safe system is a hazardous system*, but in which the *amount of hazard is small*", for example, in terms of residual risk or acceptable risk.

Further, if one assumes that the *measure of randomness* μ_1 *(in 3.5) of the occurrence of a predicted risk event R* is *some conditional* (reduced or normalized) *frequency* γ_i^*, then *the contradiction arising in the probabilistic interpretation of this measure is eliminated.* Thus, with a general approach to assessing flight safety in the SST, the risk measure μ_{IR} is universal, that is, if there is an exact probability value, it can be taken into account and introduced into the calculation. But the fact is that, as shown earlier in Chaps. 2, 3, this value cannot be found.

For example, it is sufficient to assume that a risk event is possible (for example, with *the "near-zero" probability*), then one can immediately proceed to the *ALAPA, ALAPR* [19–21] methods and FMEA, but in the form of Fuzzy Sets.

5.5 SST Application for Assessing ATS Safety Levels in the Class of Rare Events Using Classical RT and PSA Methods

The RT and SST methodology is applied on equal terms, but in accordance with the requirements of ensuring ATS properties by separate clusters of parameters reflecting the ATS production before the analysis of possible consequences from functional failures in the expected operating conditions (in order to test the significance of Prdf “heavy tails”).

Three phases of ATS production and application are established.

Stage 1 (Phase 1)—achievement of standard RT indicators on the basis of the PF without taking into account the requirements for system safety (and industrial safety).

Stage 2 (Phase 2)—providing indicators of high reliability of systems, taking into account the requirements of SST and norms of the “residual risk” for ATSs during their life cycle, based on the industrial ILS strategies by the type of samples of service systems for foreign-made aircraft or Ka-32 helicopters.

Stage 3 (Phase 3)—assessing and maintaining the level of system safety (IS) by managing risk parameters based on proactively identified risk factors and types of risks and expected risk events for selected systems.

The *action strategy in civil aviation of the Russian Federation within the “New Doctrine” (RRS) in the transition to the IATA ILS system* follows from the logic of constructing the structure of the SST module presented in Fig. 2.2 in Sect. 5.2. The *main requirements* are as follows:

- *The system produced must be highly reliable.*
- *The manufacturer of the equipment must provide quality* (RT property) such that the “residual risk” by *the probability of a risk event is not worse than* 10^{-4}–10^{-6}.

Otherwise, the service provider, for example, an airline, will not have a “risk” margin, since the airline must also ensure the significance of its risks (*norm of the “risk”*) within acceptable levels established by the state in accordance with the recommendations of ICAO amendment No. 101.

5.6 Steps to Ensure the System Safety Level for ATSs and Dual-Purpose Equipment in Terms of “Risk” During the Life Cycle of the Product

The general scheme for ensuring the flight safety level follows from the structure of the SST module (Fig. 2.2) and reflects a harmonized combination of the results of the production of technical complexes, taking into account the RT and SST requirements. Therefore, two steps of production and operation of technical complexes are distinguished that include *flight safety monitoring by factors F1 and F2.*

5.6.1 Step 1. Creation of a Highly Reliable Technical System

High reliability of the technical system is determined by the following operations.

- Provision of the standard reliability indicators on the basis of the provisions of the classical reliability theory and regulatory documents and standards (*mean time between failures, performance factors, lifetime, availability,* etc.) in accordance with the *hypothesis of the truth of events on the hypercube of the states* indicated in Fig. 5.3 and in [22–24].
- Evaluation of standard system safety indicators based on standard PSA methods (*I. Aronov, R. Islamov*)—for IS-2)—for determination of system safety indicators *by F1, F2 for IS-2*.
- Regulation (*according to the acts*) of standard reliability indicators and residual risk values (*in terms of the probability of risk events such as functional failures*) by the main critical risk factors with the most significant negative consequences.
- Constructing structural connections of reliability elements in modules of structurally complex systems, in assemblies or in the system as a whole, indicating redundant elements and redundancy management devices [19, 22].

a). Scheme variants

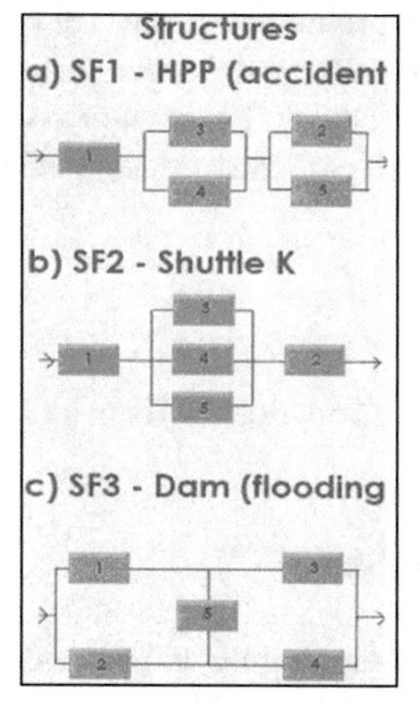

b). Element selection panel

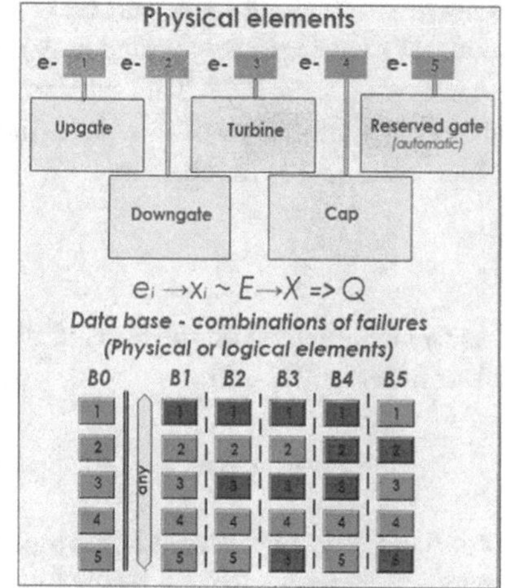

Fig. 5.3 Examples of constructing “clear” schemes for connecting reliability elements

5.6.2 Step 2. Identification of Paths Leading to a Catastrophe on the Basis of the Adopted Structural Connections of Reliability Elements

Basic positions. The residual risk $\hat{R}_0$ from (4.13) of the manufactured equipment is due to the design and production features and corresponds to the processes at the stages of system operation and the monitoring schemes for the RT indicators per typical maintenance regulations, for example, after-sales service and maintenance and repair. It is accepted, according to the basic RT provision, that "*catastrophe* (accident) occurs when the system falls into the *minimal cut set of failures*" [19, 22] or a chain of events that determine the functional failure in the form of a "*Reason chain*" [25], or a chain of logical elements of the system studied by CATS methods [26].

To search for the MCF, it is suggested in the SST that the direct search method should be used for combinations of events *like "failures"* in event trees (by RT, CATS, FMEA methods), but, unlike the RT, *without calculating "failure probabilities"*, since *the probability value determines only a measure of random occurrence of an "accident", but does not affect the consequences or physical conditions of a particular "accident"* [2, 10, 15].

According to the provision on the existence of "Step 1" ("On a highly reliable system"), ensuring the reliability of particular AE production causes the occurrence (i.e., "introduces") the inevitable (mandatory) risk $\hat{R}_0$ by the probability P_{**0} of a risk event R not better than 10^{-6}. This substantiates the SST provision in Chap. 2 of the book about the significance of this level (inevitable "from above") with the risk of the first kind from (3.12) in the form:

$$\mu_{1P0} = \mu_{P0**} = 10^{-6}.$$

In fact, as can be assumed on the basis of the information in Chap. 1 (Table 1.7), for the nuclear industry [27–32],

$$\mu_{P0*} > \mu_{P0**} = 10^{-3}\text{–}10^{-5}\left(\text{not } 10^{-6}\right).$$

It is worth mentioning here what is the aim of developing an after-sales service strategy, including maintenance and repair for "Boeing" or "Airbus" aircraft in civil aviation of the Russian Federation with the use of the MEL and MMEL programs [8, 12]. This strategy implies that new aircraft can be put after production into operation with a residual risk $\mu_{P0*} \gg \mu_{P0**}$ of 10^{-6}, for example, 10^{-3}–10^{-5}. But in new ILS technologies (maintenance and repair), this turns out to be non-essential, since with the service strategy considering the actual state and due to the current minimum redundancy of ATS elements per MMEL, the actual value of the residual risk with the measure μ_{P0*} will be acceptable and approach the value of 10^{-6}.

Stage 2 Procedures The procedures for this stage (phase or some production steps) are as follows.

1. For simplicity of calculations of the first approximation, constructive schemes of connections of the system reliability elements are developed in the form of "clear" schemes—within t classical RT approaches, as shown in Fig. 5.3a, b.
2. From the ATS passport data, a list (set) of the minimal cut sets of failures (MCFs) in the AE is identified with the indication, if possible, of conditional probabilities of the occurrence of the specified combinations of failures in the identified minimal cut sets of failures according to known methods ("failure tree").

This is necessary for assessing the degree of approximation of the randomness indicator for a risk event (functional failure) at the level of 10^{-4}–10^{-6} in terms of the FRNS.

3. A hierarchical set of functional failures of the system (AT) is determined that lead to catastrophes (accidents) of a given rank with a ranked set of possible predicted negative consequences (damages).
4. A database on possible damages for each of the functional failures is compiled; a list of hidden threats and risk factors is determined and classified that should be taken into account in reason events scenario chains.
5. The search (detection) of possible paths to a catastrophe is performed in the form of some scenarios [8, 12] ("Canada") by the "event trees" method (FMEA) or by the J. Reason chain method, for example, on the basis of programmable complexes [33, 34] known in the RT or developed by OJSC "Aviatekhpriemka" [6, 35].

Examples of the results obtained with the help of the listed programs (developed by "Aviatekhpriemka") are shown in Fig. 5.4a, b. Here, graphs are given for searching paths and the paths themselves to a catastrophe in the form of "cones of risk" and *"X-mas tree"* (ADS)—a FMEA analog in the ADS forme.

The *"X-mas tree"* is shown in Fig. 5.5a, b (below).

6. Based on the filter method used in J. Reason chains and the database with ranked damages, the methods of corrective risk management in ATSs are selected with identified risk factors.
7. An alternative action strategy is selected: No. 1 "Hit or miss"—according to NASA; No. 2—"With risk management"—according to FORAS (NASA), ACARS [36], MEL [37].

This is necessary to proactively improve the system safety and upgrade the ATS lines of protection from the risk event R_0 with the indicator of the significance of the risk $\hat{\mathrm{R}}_0$ compared with the indicator $\hat{R}_{0*}$ of "acceptable risk".

EXAMPLE SBM (5)

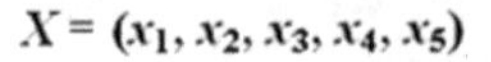

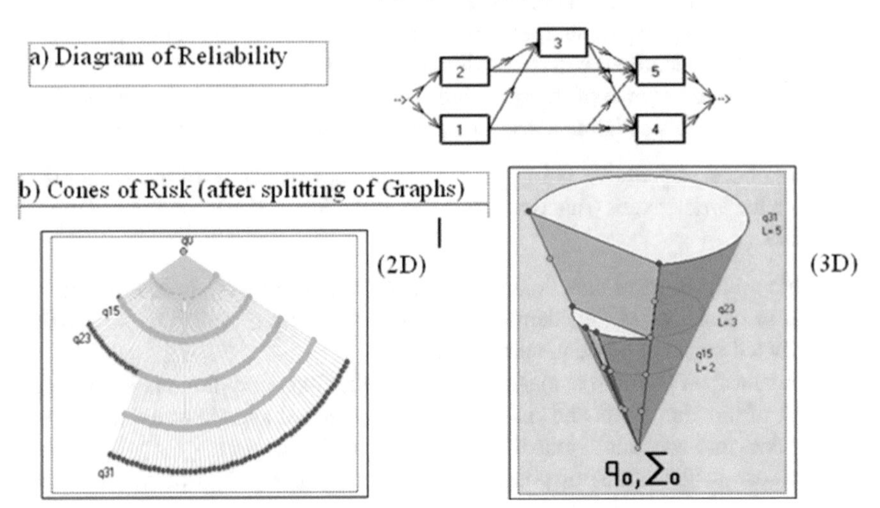

Fig. 5.4 "Cones of risks" for a given diagram of five elements

5.6.3 *Formalized Models of System Structures Taking into Account Possible Failures Based on Models of the "Hypercube of Truth"*

The S system model is adopted in the form proposed in Chap. 3:

$$S = S(X, Q, \Gamma Q | \Sigma_0),$$

where Σ_0 denotes the same conditions and models of the existence of system elements and technology for their application and maintenance (according to the RT positions from Chap. 1) as in Sect. 4.3.2. There is a clear universal set of attributes:

$$\Gamma Q : Q \to Q,$$

where Q is a countable set of discrete states of the system, ΓQ is a graph of changes of discrete states in the S system.

But now, according to the SST, the following is introduced

$$Q = \{q_i | \Sigma_0\}, \quad i = 1, 2, \ldots, n.$$

This allows solving the rare events problem within the RRS.

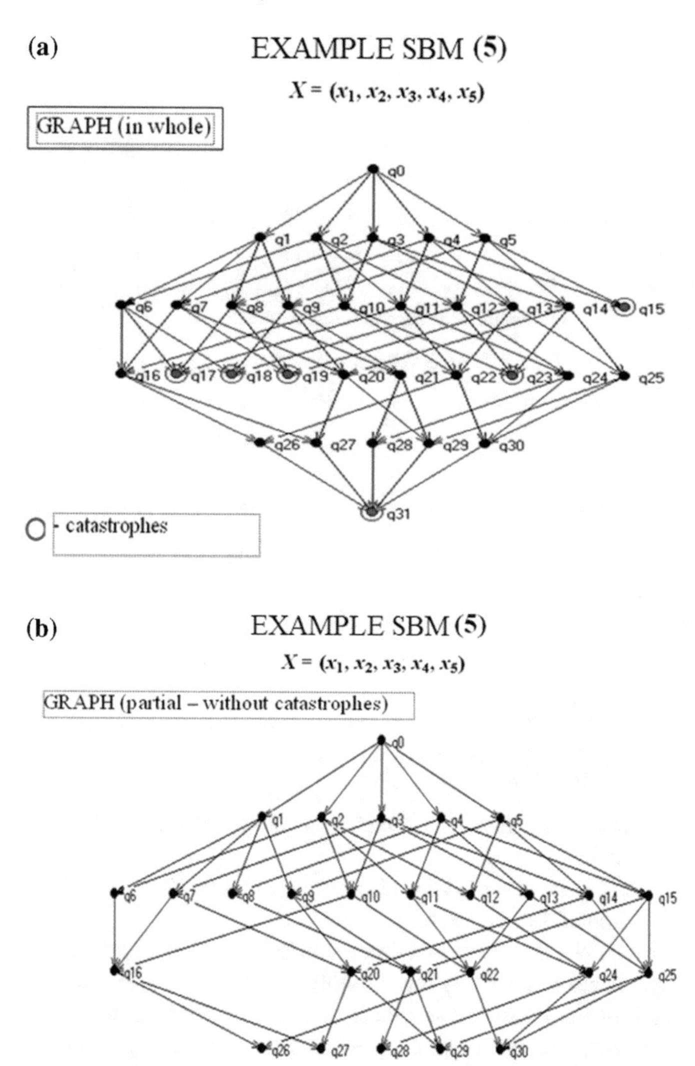

Fig. 5.5 **a** "Risk" ("risk combinations of events", "X-mas tree" program). **b** "Chances" (combinations of states and paths—"without catastrophes", "X-mas tree" program)

Examples of constructing "clear" schemes for the case $n = 5$ are shown in Fig. 5.3. These schemes, as known "in production", are included in the databases (DATA BASE), in the safety systems of technical complexes described in Sect. 4.3.

These schemes can also be used in assessing the risks of adverse consequences using the SST approach from the domain of fuzzy subsets when studying the properties of rare events [38]. In Fig. 5.3 from Sect. 5.6.2, the following designations have been adopted:

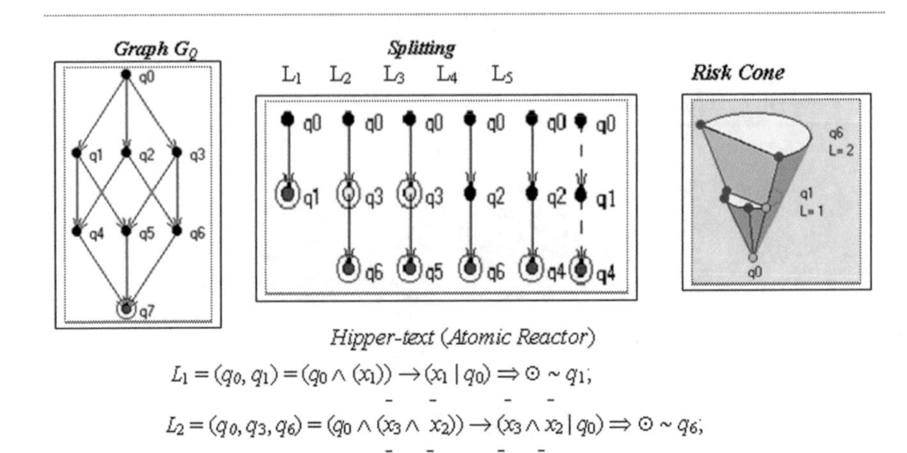

Fig. 5.6 Diagram and risk cofone (atomic reactor)

- SF1—submarine [39],
- SF2—Shuttle "Columbia" [16],
- SF3—hydroelectric power station [40].

The development of "paths to a catastrophe" is given in Fig. 5.5a ("Risks") and Fig. 5.5b ("Chances").

Similar results were obtained for the system *with* $n = 3$; here, J. Reason chains are shown in a logical form (Fig. 5.5).

Figure 5.6b shows the enlarged combinatorial part of the general algorithm for constructing J. Reason chains by the example of a set of elements with $n = 3$. The practical value of the presented results is that, firstly, the Reason's "Swiss cheese model" [12] is described mathematically, which for the time being nobody did (except attempts at CATS). Secondly, the reason chains denoted by symbols L_i are expressed in terms of the "event tree" through the combinatorics of the elements and logical connections in the form of "conjunctions" of elements in each chain. The original object is a graph G_0. The chain is obtained as a result of "taking a knitting needle out of cheese" and "reading marks from traces of cheese layers from the knitting needle". Further, a "risk cone" and analytical descriptions of the chain structure are given.

Further (in Sect. 3.10), an example of such an algorithm is shown that is used in the problem of assessing risks of the occurrence of aircraft accidents during a flight along the given route.

5.7 Model of Estimation of the Counterfeit Influence on ATS Safety in Fuzzy Sets

In the classical RT, methods have been developed to ensure the **quality of systems** from *the standpoint of consumer demand* only on the basis of the probabilistic concept of determining certain standard indicators. There is a misconception presented by the formal transfer of calculated probabilistic reliability methods to the safety assessment process. The *justification of structure and traditional features of the safety problem* under consideration follow from this, taking into account the manifestation of "***counterfeit***" properties. New approaches are needed, for example, the RRS scheme from this section of the book.

The ***essence of scientific results consists*** in the development of a *counterfeit model* in the form of an "aging element" with a *two-parametric function λ (lambda)*, in the application of new maintenance and repair technologies, *diagnostic algorithms for the state of third-generation products*. The problem differs from the known ones in that the hazard of failure from this non-authentic element is introduced with *unknown information about the "reference point" (datum)*, but in the form of an a priori density of probability distribution at this point.

Such a scheme is proposed for the first time in this book for the case when the initial system (in terms of reliability) reduces to a parametric system with a parameter from which the "aging" interval is measured. In this scheme, it is necessary to proceed to the concepts of "risks" of the occurrence of negative consequences in the modern interpretation and also to find ways to compensate for the consequences.

The key point is the ***development of "counterfeit" models. After that, the effect of counterfeit properties on safety indicators is assessed*** in accordance with the ICAO methodology on the basis of the risk concept.

It is assumed that the failure hazard coefficient λ of some non-authentic element is specified as for an aging element by means of a non-stationary λ function in the form of λ_{na}, but with an additive:

$$\lambda \to \lambda_{\text{na}}(t) = \lambda_0 + \Delta\lambda(t), t \in [0, T).$$

If it is traditionally assumed now that $\lambda_0 = \text{const}$ and $\Delta\lambda(t)$ are known, then this counterfeit model reflects the "acceleration" of the failure process and the "deterioration" of a number of standard quality indicators φ_κ, in particular, the non-failure indicator. From this, it follows trivially that it will be necessary to substantially increase the volume of spare parts; this explains the uneconomical nature of such principles of ensuring the reliability of systems. This conclusion was formulated in [41] by E. Barzilovich, and a scheme was proposed for choosing the optimal stock of spare parts on the basis of the minimax method. ***But in practice the situation is more complicated***, as noted above.

Another incorrectness is the average time to a catastrophe (accident), which cannot even be calculated in order to evaluate the "lambda". A small number of catastrophes with "Airbus"–"Boeing" aircraft with extensive damage do not provide the necessary information, especially since the number of possible catastrophes is not known in advance, so there must be other models for assessing the hazard level and searching for ways to improve the quality of systems.

The scheme of parametrization of standard indicators of ATS reliability allows for correct solutions of the formulated task. ***One has to assume additionally that the initial value of*** λ_0 ***is unknown in advance, and the moment*** τ_0 ***as the datum of the failure rate for the non-authentic element is random, because the background of the element life cycle is also unknown. Therefore, the following is proposed*** [14, 42]:

$$\lambda \to \tilde{\lambda}_{\text{na}} = \tilde{\lambda}_0 + \Delta\lambda(t - \tau_0) = f_\lambda(\lambda_0, \tau_0).$$

This means that the hazard coefficient $\tilde{\lambda}_{\text{na}}$ is a two-parameter function. Formally, this can be taken into account in the calculations, since this does not lead to either theoretical or methodological difficulties. Here, a concept of integral standard quality indicators $\hat{\phi}_c$ is introduced:

$$\hat{\phi}_c = f\left(\left\{\varphi c(\tilde{\lambda}_0, \tau_0) | \omega_{\text{na}} = (x, y)\right\}\right),$$

where $\omega_{\text{na}} = (x, y)$ is some now already two-dimensional probability density function for parameters $\tilde{\lambda}_0 \sim x, \tau_0 \sim y$ with respect to arguments *x, y*, taking into account known constraints on τ_0, a random variable of positive terms in the form of "time". This model is more believable, as the real counterfeit hazard factors are taken into account.

The conclusions from this are the same as [41], but even more pessimistic, namely in addition to increasing the volume of spare parts, it is necessary to reserve almost all the elements of the ATS structure if the list of counterfeit products is unknown in advance, and safety and reliability must be guaranteed.

Thus, in the ATS systems of old generations, the *problem of counterfeit in theory cannot be solved in any way on the basis of the methods of the classical reliability theory; only the trivial approach remains* that is to increase the volume of spare parts and reduce the level of safety. All this is due to the uncertainty of the information.

There are only two methods of solution: "Monitoring the path of each ATS element with the help of tags or product identifiers", which has been done in civil aviation of the Russian Federation for many years, and this "saved" civil aviation. Another option is the method *proposed in this book* for the transition to the introduction of the *principle of diagnostics of failures by the state* with compensation for increasing risks based on the MEL program. At the same time, the increase in the volume of spare parts is controlled by the indicators of the risk of function loss, as it was done for foreign-made aircraft. Obviously, it is necessary to develop a new maintenance and repair strategy and standards, taking into account the requirements for safety indicators in units of integral risk. But the concept of integral risk is introduced only in the RRS.

It can be noted that the MMEL manual for AN-148 aircraft [37] specifies the need to use spare parts taking into account the calculation of risks, but the necessary methodology is not provided. (In similar tasks, the SST tools developed in the RRS based on the ICAO recommendations can be useful).

5.8 Analysis of the Combinatorics of HF Characteristics with the SHEL Interface

The goal of the section is to show that there is no "Human factors problem" [43] usually considered in civil aviation (V. Kozlov), since this problem can be solved simply by dividing its content into two parts based on the recommendations of the new RRS doctrine (adopted in this book). To do this, it is enough to use models of calculating the "risks of negative consequences" for ATSs in the new RRS interpretation. The bottom line is that it is sufficient to preliminarily determine the entire set of combinations of states in the SHEL, the number of which is "always countable" and can theoretically be found and numbered.

5.8.1 Statement of the Problem and the Solution Scheme

The conventional name "human factor" [43] only means a set of indicators of the negative impact of certain factors of human activity (for example, operators in the cockpit, controllers in the ATM system) on safety processes in civil aviation.

These features, as is known [43], are the subject of research conducted by aeronautical experts and first of all by psychologists. It is believed that mental activity (the work of the "brain") is difficult to formalize [44].

SHEL is the ICAO (NASA) designation of the aircraft subsystems interface [12] required to ensure a safe flight: "human (operator)", "technic (equipment)", "control program", "external environment" (surrounding conditions) (Fig. 5.7a,b).

The (negative) in-flight HF manifestation is considered one of the causes of the "emergency situations" and catastrophes with civil aircraft. (One of the experts in the field of studying HF in civil aviation is Professor A. Kozlov from JSC Aeroflot, whose research focuses on studying the "thinking activity of pilots" during the flight, and other issues, which is extremely important, but also difficult to study).

From the RRS positions, the known approaches to solving the HF problem using probabilistic and psychomotor models of the operators' behavior and ways to compensate for the consequences considering the "learning" function are not sufficiently informative.

The proposed method for solving the HF problem (according to the RRS) is quite simple (in two parts), as noted above:

- identification (separation) and classification of all types of interaction for "operators and equipment" based on the SHEL; the obtained set will always be countable and clear, conforming to the "Boolean lattice" hypothesis and be completely determined by all documents, instructions, and regulations for a particular ATS type;
- a description of all possible clear states on the basis of graphs (Fig. 5.7b) with clear possible transitions;
- determination of a set of activity efficiency and quality indicators, which are always fuzzy (the psychologists' practice includes terms like "trained", "not trained", "active", "irresolute", etc.).

Part 1 concerns the "capabilities" and properties of a functioning system that are always clear. *Part 2* concerns measures of opportunities that are always fuzzy and reflect the motives for the operator's choice of "suitable" actions that depend on both the professional skills and the psycho-emotional state of the operators.

On this basis, different combinatorial schemes can be a priori analyzed for automating procedures and algorithms for analyzing features of functionality and consequences with an assessment of catastrophe risks. Appropriate databases can be created that should be standard. But the risks of catastrophes in such situations are estimated quite properly only with the help of the SST tools (without probabilistic characteristics, which mainly "do not exist" and will not be needed for fuzzy expert methods for assessing the risks of negative events).

5.8.2 Coding of SHEL States

The initial diagram in Fig. 5.7a,b is transformed into matrices of codes presented below on the basis of combinatorial analysis of events.

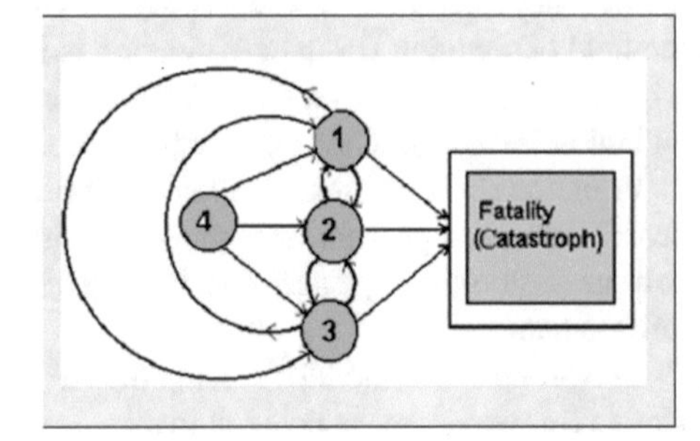

Fig. 5.7 Interpretation of SHEL-concept

The baseline is the following interpretation of the SHEL interface. For a diagram as in Fig. 5.7a, a functional space of discrete states q_i is considered.

A discrete state is a set of four elements a_{ij} (tetrad):

$$\text{Values of } a_{ij} = (0 \text{ or } 1),\ q_i = \{a_1(i), a_2(i), a_3(i), a_4(i)\}, \quad i = 1, 16,\ j = 1, 4.$$

Here, only 64 a_{ij} elements are identified—for 16 states of digital codes in case of "failure" in links of the SHEL system in processes leading to a catastrophe. The table of SHEL system codes shown in Fig. 5.8 is as follows:

$$M_d = f(n_s, MS_{|\text{SHEL}}, N_d); \quad n_s = 4; \quad MS_{|\text{SHEL}} = M_s(4), \quad N_d = 2^4 = 16.$$

A database based on statistics can be compiled for each code, which is necessary for automated analysis of emergencies and identifying the causes of incidents, including the design ATS features. Thus, with the help of the SST recommendations, the first part of the HF problem can be solved (i.e., an assessment is made of what can be objective in a given ATS (with a trained operator and approved flight standards).

Further, the second task can be solved—the search and analysis of paths leading to a catastrophe and the identification of risks characterizing the level of hazard.

The second problem can be solved with the algorithm is used to analyze the combinatorics of events shown in Fig. 5.4 (above, see Sect. 5.6).

For this purpose, using the SHEL code table (Fig. 5.8), a matrix of system state-to-state transitions is built, according to the SHEL interface (Fig. 5.7a), for a given transition graph in the same way as in Fig. 5.7b.

The transition matrix is given in the form of an adjacency matrix [45] in Fig. 5.9. Here, graphs can be obtained either by the "X-mas tree" method (ADS), as in Fig. 5.5a,b or by the standard FMEA procedure. Then "risk cones" and chains of possible scenarios for the development of hazardous and "special" in-flight situations are identified. It should be noted that formally the FMEA is an approach to analyze

Таблица кодов системы SHEL

$M_d = f(n_s, Ms_{|SHEL}, N_d)$; $n_s = 4$; $Ms_{|SHEL} = M_s(4)$, $N_d = 2^4 = 16$

q_i	q_0	q_1	q_2	q_3	q_4	q_5	q_6	q_7	q_8	q_9	q_{10}	q_{11}	q_{12}	q_{13}	q_{14}	q_{15}
d_z / α_r	d_0	d_9	d_5	d_3	d_2	d_{13}	d_{11}	d_{10}	d_7	d_6	d_4	d_{15}	d_{14}	d_{12}	d_8	d_{16}
α_1~S	0	1	0	0	0	1	1	1	0	0	0	1	1	1	0	1
α_1~H	0	0	1	0	0	1	0	0	1	1	0	1	1	0	1	1
α_1~E	0	0	0	1	0	0	1	0	1	0	1	1	0	1	1	1
α_1~L	0	0	0	0	1	0	0	1	0	1	1	0	1	1	1	1

S – **Surrounding** **H** – **Hardware** **E** – **Tngineering** **L** - **Life**

Fig. 5.8 Example (SHEL)

MATRIX (basical)
OF POSSIBLE TRANSITION
based on set of discrete states for SHELL -system

L $\{i = 1, 16\}$

H $\{0, 1\}$

	1	2	3	4	5	6	7	8	9	10	11	12	13	14	15	16
1	//1//	1	1	1	1	1	0	0	1	1	0	0	0	0	0	0
2	1	//1//	0	1	0	1	0	0	0	1	0	0	0	0	0	0
3	1	0	//1//	1	1	0	1	0	0	0	1	0	1	0	0	0
4	1	0	1	//1//	0	0	0	1	0	0	0	1	0	0	0	0
5	1	0	1	1	//1//	1	1	0	0	0	0	0	1	0	0	0
6	0	1	0	0	1	//1//	1	1	0	0	0	0	1	0	0	0
7	0	0	1	0	1	0	//1//	1	0	0	0	0	0	0	1	0
8	0	0	0	0	0	1	0	//1//	0	0	0	0	0	0	0	1
9	1	0	0	1	0	0	0	0	//1//	1	1	0	1	0	0	0
10	0	0	0	0	0	0	1	0	1	//1//	0	1	0	1	0	0
11	0	0	1	0	0	0	0	0	1	0	//1//	1	0	0	1	0
12	0	0	0	1	0	0	0	0	0	1	1	//1//	0	0	0	1
13	0	0	0	1	1	0	0	0	1	0	0	0	//1//	1	1	0
14	0	0	0	0	0	1	0	0	0	1	0	0	1	//1//	0	1
15	0	0	0	0	0	0	1	0	0	0	1	0	1	0	//1//	1
16	0	0	0	0	0	0	0	1	0	0	0	1	0	1	1	//1//

$\{1\}$- general diagonal, **L** $\{i = \overline{1,16}\}$ - initial states (upper string)
H $\{0, 1 | \overline{1, 16}\}$- current states and point to transition

Fig. 5.9 Basic matrix for SHEL-system

the "event tree", but without adaptation to the SHEL interface and with evaluation of only "probabilistic" indicators of the state criticality. The latter is completely unacceptable, uninformative and has no practical meaning. Therefore, a new scheme is provided for solving the problem posed in Sect. 5.8.1.

5.8.3 Risk Assessment Based on the SST (RRS) Algorithms

In this task, J. Reason chains (in the form of scenarios) are constructed on the basis of SHEL databases (DBs) (actively or proactively), which is done in the corresponding flight safety management system (SMS). Risks can be determined using the analysis software for "event trees" used in the RT in the form of FMEA (international standard). For this purpose, the scheme of Fig. 5.5 is used with the selection of possible scenarios of events in a logical form.

The measure of assessing the possibility that the "operator" "suddenly violates" flight standards belongs to another area of research. The key point is that the "equipment" allows only correct actions prescribed in instructions and AFMs. But the possible "harm" (damage) from errors in the operators' behavior is proposed to be evaluated on the basis of risk models and hazard models within the countable combinatorics of events determined with the transition graph in Fig. 5.4.

All variants of system state-to-state transitions are known in advance and countable; the combinatorics is defined and fixed in databases, since the proposed models are determined on discrete states, the set of which is also countable. This is shown geometrically in the example of the SHEL interface specified by the transition matrix in Fig. 5.9, where the logical symbols "0" and "1" are indicated, which follow from the Boolean lattice of the system with the given structure.

(Note. The authors presented here copies of the originals of their schemes, which were discussed at international meetings).

5.9 Layers of J. Reason Chains for Proactive Determination of the Preconditions for Aircraft Accidents in Flights

The benefits and necessity of applying the principle of constructing predicted hazard models in aircraft flights are shown on the basis of the calculation of risks as an amount or a measure of hazard in the SST and RRS tools. J. Reason's method is widely discussed in ICAO standards in the form of a "Swiss cheese diagram" [12]. But over the past 10–15 years, nothing but pictures is discussed at various meetings (presentations). In this book for the first time, the base of combinatorial methods is introduced for the "cheese model". The essence of J. Reason chains was discussed in Chaps. 3 and 4 of the book. In particular, a connection was established with the FMEA program, which, by the way, is the basis for the "IL Corporation" [46] (paper by M. Neymark et al.). Here, the interpretation of J. Reason chains in the form of layers is used. An example is given of the preflight preparation of crews for flight plan assignments.

The *layer method* provides *structuring of a system* of random *factors* that can lead to a serious negative result in the analyzed situation and can influence the occurrence of a hazardous event or a hazardous situation in flight.

The *aircraft flight* is represented as a *conditional trajectory* that passes *through all levels of layers one by one in the J. Reason chain*: from the initial and to some last one, where a hazardous event can occur, if preconditions for an accident have occurred in one of the preceding layers.

Any risk events can be *modeled* according to the J. Reason chain scheme. By the way, in the Canadian version of the SMM, only the concept of "scenario" is used, without reference to the author.

The designations of the layers are as follows: of the flight operations directors (No. 1); control along the flight route (No. 2); prerequisites (3'—constant and random errors No. 3); hazardous actions (No. 4); various outcomes (No. 5); the "window" of opportunities (No. 6).

Despite the complexity of the chain and multiple "layers", the risks (integral ones in the terminology of Chap. 3) are evaluated only with the help of the "risk assessment matrix" from the SMM (Annex 19). This is of particular practical value, since the construction of a detailed probability space is not required, and one can limit themselves to conditional binary partitions of the outcome space.

The algorithm scheme is shown in Fig. 5.10.

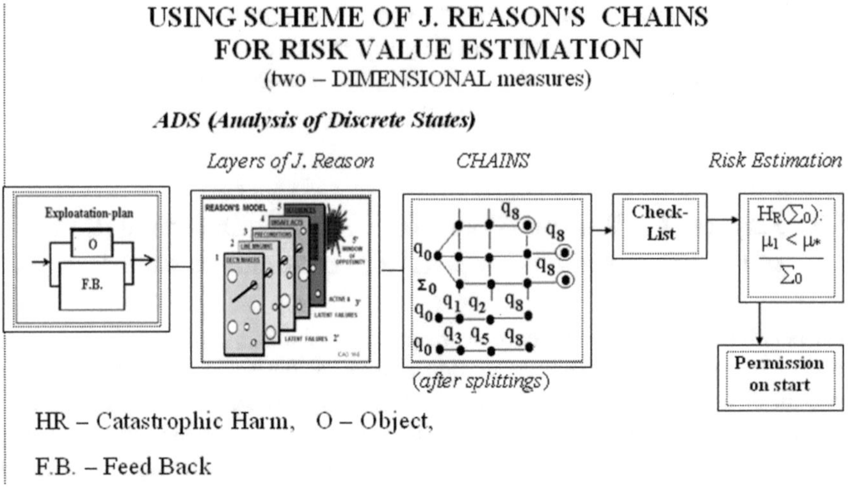

Fig. 5.10 Scheme of risk value estimation, based on chains of reason

All paths based on the combinatorial principle can be "unfolded" using the ADS ("X-mas tree") computer program, which was demonstrated many times by the authors of the book at various meetings.

So, the "crew" and the "aircraft" cannot influence the weather, as "weather" is an external adverse factor (risk factor). But, according to common sense, as reflected in the algebra of clear and fuzzy logic and in the combinatorics of a countable set of events, the aircraft can avoid the factor and not accept the risk. Thus, in the problem under consideration, the PSA methods may not be used.

5.10 "Risk and Vulnerability Points, Vulnerability Intervals on ATC Trajectories with the "Vectoring" Method" (ICAO and Annex 19)

Background ICAO ("Annex 19") announced the program for creating SMSs on the following basis [19]: to provide SMS requirements (ARMS-ECAST type), but with the solution of the "rare events" problem (per ICAO) by creating databases on NASA samples (*principles and approaches*), taking into account the limited volume of statistics and with the uncertainty of the possibility measure (or *randomness, in simple situations*) for the occurrence of rare events; it is recommended to apply "preventive" (proactive) management of the STS state taking into account risk factors on the basis of ICAO algorithms with limited statistics (with uncertainty of the measure of rare event occurrence); ensure [11] "Monitoring the state" of flight safety.

Indeed, it is impossible and unprofitable to provide absolute reliability with mathematical and technical means, since any increase in reliability is related to the costs of production, which should be, to a certain extent, optimal, based on justified criteria of optimality and efficiency [19].

Problem Description

Schemes are provided for solving problems within the ***Fuzzy Sets*** approach on the issue "Designing SMSs for ATC systems" **on the basis of *"fuzzy domination of priorities on aircraft flight trajectories"***. The main point in the problem is to find a scheme for automating typical procedures for proactive aircraft control in airspace based on the methods of "preventing hazardous proximity of aircraft by correcting flight-level parameters based on ICAO SARPs".

Here the problems are the same as for the aircraft operation based on the criterion of ensuring a specified level of acceptable risk of situations arising with the "near-zero probability".

The purpose of the presentation is to substantiate the possibility of solving a well-known Reich's problem adopted in all ICAO manuals on flight safety in ATC, in particular for flights in RVSM modes.

It is necessary to substantiate the application of the new risk formula arising from the classifier in Table 3.1 (above). It is also necessary to show how the problem was solved by M. Fujita, as described in the report on the ICAO grant implementation results.

These issues are solved using the method of describing the kinematic trajectories of the aircraft movement based on the "*vectoring*" principle adopted in the relevant ICAO manuals.

Models of aircraft traffic patterns are interpreted based on the EU Eurocontrol recommendations (the European Union) are used. On this basis, it is possible to give a reasonable interpretation of such concepts as *"risk points"*, *"vulnerability points"*, *"vulnerability windows, or vulnerability intervals"*. At the same time, the unity of the "risk-oriented" approach of ICAO (NASA) on the basis of (Doc 9859-AN/460-2007) is proved in the problem of ensuring safety of aviation activities in various domains, including the ATC domain (ATM in the modern interpretation). In the ATM domain, the provisions of Annex 19 are not fully used and are interpreted with some differences. The main result is a short declaration that some issues regarding the SMS are such that "it is necessary to fulfill them".

For "Risks" Section

The structure of SMS databases and specific processing peculiarities are insufficiently substantiated in light of the rare events problem per ICAO in the ATC domain. Special tools such as "*risk assessment matrices*" are recommended by ICAO as a tool to compensate for uncertainty in a "deadlock" situation in the absence of statistics. As an example, a promising ATC warning system is considered as an example, such as the short-term conflict alert system (STCA—short-term conflict alert, the third level of collision avoidance systems).

General Remarks and Some Methodological Misperceptions

In Eurocontrol documents, particularly in those regarding the STCA, the concepts "probability" and "possibility" are confused.

The concept of "*risk probability*" does not match the definition of "*risk*" per ICAO. (*If risk is a probability*, as is customary in old Russian standards, then this concept is more than *ill-posedness*.) This remark was discussed in the previous sections and was reflected in the classifier of uncertainty types in Table 3.1.

Probability refers only to events. Probability cannot be guessed: one can only accurately and precisely calculate if there are corresponding distribution functions for certain quantities.

Typical interpretations of the concepts "proactive management", "scenarios", "sources of hazard", and "risk matrix" have long been designated as SMS and DB objects. Threats are the first element in the algorithm for identifying hazards and risks using the NASA formula.

Intelligent support for controllers' actions who must make decisions in conflict situations and be familiar with the basics of the "risks and chances weighing" methods using Fuzzy Sets methodological positions within fuzzy logic, as follows from the NASA and the FAA recommendations. This reflects the real "human activity" when operating with fuzzy concepts, which was the impetus for the creation of the Fuzzy Set concept.

In case of situational analysis of conflict situations in the ATM system, proactive prediction of catastrophic events is virtually always performed according to the methodology for identifying "vulnerability points and fuzzy vulnerability windows". Scenarios of events are predicted in real time by analyzing each (any) arising threat (or if signs of incidents are present) with an assessment of the results criticality without applying probabilistic indicators. Although the TLS indicator is used in view of the absence of any statistics on rare events, the methods of probability theory are used traditionally.

Eurocontrol risk parameters in conflict situations recorded in the STCA

A conflict situation is a situation of an aircraft encounter characterized by a ***predictable*** violation of the specified intervals for the aircraft separation.

There are four categories of hazardous aircraft encounters depending on the degree of violation of the safe separation intervals: the time-to-alert and the point of risk, etc.

The time-to-alert parameter is measured as the time from the moment the STCA triggers to the point of risk (PoR). It depends on how the point of risk is defined.

The concept of a "point of risk" is introduced to determine the moment to which the time-to-alert is measured.

The point of risk can be different for different STCA implementations and depends on the implementation logic used in each individual system. A point belonging to a real or predicted flight path of an aircraft (aircraft) may be considered a point of risk and measured by distance in time, space, or a combination of these values depending on the specific implementation. Here it is necessary to amend the definitions taking into account the provisions of Annex 19.

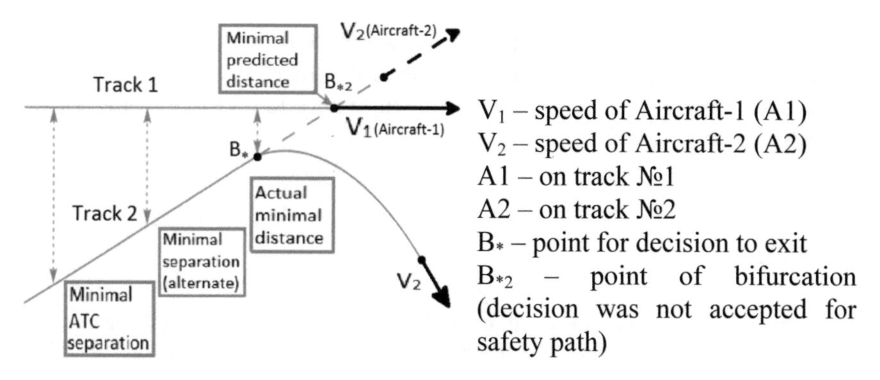

Fig. 5.11 Options for determining the point of risk for the STCA

In Eurocontrol, for the STCA, the point of risk can be defined as the closest point of approach (CPA) or the boundary of violation of any specific criteria (for example, established separation standards). It should be noted that if the point of risk is defined as a CPA, then the time-to-alert should be longer to ensure the same level of safety than if the point of risk is defined as the boundary of violation of the separation standards. Figure 5.11 shows several options for determining the point of risk that can be used in the STCA system.

The presented scheme includes harmoniously only the new positions of the axiomatics of risk models developed in this book.

To ensure effective operation, the STCA system must be implemented as part of the ATC AS (automatic system) using high-quality surveillance and processing information with high resolution or a prohibited flight level for more reliable conflict detection;

To ensure effective operation, the STCA system must be implemented as part of the ATC AS (automatic system) using high-quality surveillance and processing information with high update rate.

Similar proposals found application in the **national standard *"Monitoring, Navigation, Communication, and Automation aids of the ATM in Civil Aviation of the Russian Federation. Technical specifications"*** *developed by the Almaz-Antey Corporation (Branch "Aeronautical Research Institute") Snt.Peterburg: 2015.*

5.11 Conclusions—5

1. It is shown that the tools for assessing the significance of risks as a measure of the amount of hazard in the states of the systems under study allow solving practically all the issues of the rare events problem on the basis of the Fuzzy Sets methodology without using probabilistic indicators.

2. The hazard models proposed by ICAO, such as SHEL, HF, J. Reason chains, FMEA scenarios, CATS, and others to describe the properties of aviation and technical systems can be correctly constructed using the SST tools defined in the RRS doctrine.

References

1. Kuklev EA (2005) Estimation of catastrophe risks in highly reliable systems. In: Materials of the 13th International conference "Problems of complex system safety management. RAS ICS. Moscow: 55 p (in Russian)
2. Kuklev EA (2005) Decision making in systems based on management of possible risks of undesirable consequences. In: Materials of the 13th International conference "Problems of complex system safety management". RAS ICS. Moscow, 87 p (in Russian)
3. Standard for the assessment of major aviation risks (2012) Version 4 (Threats). (Raw material industries). Flight Safety Foundation. Melbourne (Australia) (in Russian)
4. Gipich GN, Evdokimov VG, Kuklev EA (2010) Methodical provisions of the classifier "SMS terms and definitions" in the field of flight safety management. Collection of papers of the NTC "Rostekhregulirovanie. no. 2. Moscow (in Russian)
5. Gipich GN, Evdokimov VG (2011) The concept in the field of standardization of principles and procedures for the development of a national version of the system and manual for safety management in the production of aircraft in the UAC at the state level in the Russian Federation. Report (printed) in the "UAC Bulletin" no. 2 1. Moscow, pp 65–70 (in press) (in Russian)
6. Gipich GN, Evdokimov VG (2009) Principles of a unified approach to assessing complex system safety based on indicators of risks. Collection of papers. RAS CC "System Safety", no. 2, Moscow, pp 112–120 (in Russian)
7. SMM (2009) Doc. 9859-A/N460. ICAO
8. SMS & B-RSA (2008) Boeing, 2012
9. Amer MY (2012) 10 Things you should know about safety management systems (SMS). SM ICG, Washington
10. Evdokimov VG, Grushchanskiy VA, Kavtaradze EA (2008) On the system safety of information support for the development of complex social systems. Fundamental problems of system Safety: Collection of papers. RAS Dorodnitsyn CC, Moscow: University Book, pp 67–77. 14 (in Russian)
11. Annex 19: C-WP/13935 ANC report (2013), based on AN-WP/ 8680 (Find) Review of the Air Navigation Commission, Montreal, Canada
12. SMM (Safety Management Manual) (2012) Doc 9859_AN474—Doc FAA
13. Probabilistic Risk Assessment Procedures for NASA Managers and Practitioners—Office of Safety and Mission Assurance NASA. Washington, DC 20546, August 2002. (Version 2/2)
14. Kuklev EA, Evdokimov VG (2001) Predicting the safety level for aviation systems based on risk models. J TR, 2(45): 51–53 (in Russian)
15. Evdokimov VG, Gipich GN (2006) Some issues in the methodology of assessment and prediction of air transport risks. In: The 5th International conference "aviation and astronautic science 2006", abstracts, Moscow, pp 75–76 (in Russian)
16. Novozhilov AB, Neymark MS, Cesarskiy LG (2003) Flight safety (Concept and technology). Mechanical engineering. Moscow, 140 p (in Russian)
17. Risk management—Vocabulary. ISO Guide 73: 2009 (E/F). BSI 2009
18. GOST 23743-88 (1984) Aircraft. Nomenclature of the Flight Safety, Reliability, Testability and Maintainability Indices (in Russian)
19. Aronov IZ et al (2009) Reliability and safety of technical systems. Moscow (in Russian)
20. Savchuk VP (1989) Bayesian methods of static evaluation of reliability of technical facilities. Nauka, Moscow (in Russian)

21. Evdokimov VG (2012) Integrated safety management system for aviation activities based on ICAO standards and recommended practices. J TR. 2(45): 54–57 (in Russian)
22. Ryabinin IA (1997) Reliability, survivability and safety of ship electric power systems. –Kuznetsov Naval Academy, St. Petersburg (in Russian)
23. Orlovskiy SA (1981) Problems of decision making with fuzzy source information. “Science” FM, Moscow (in Russian)
24. Rybin VV (2007) Fundamentals of the fuzzy sets theory and fuzzy logic. Study guide. STU Moscow State Aviation Institute. Moscow, 95 p (in Russian)
25. Evdokimov VG (2010) Modern approaches to safety management based on the risk theory. International Aviation and Space Magazine “AviaSouz”. NQ 5/6 (33) October–December (in Russian)
26. CATS (Casual Aviation Technical Systems) (2012) Simulation of cause-effect relations in aviation systems on the basis of risk assessment. Research of the Air Accident Investigation Commission (Netherlands). ICAO (in Russian)
27. Shvyryaev YV (1992) Probabilistic analysis of nuclear power plant safety—method. Kurchatov IAE, Moscow (in Russian)
28. Putin VV Russian nuclear power plants are reliable (in Russian)
29. Kirienko AN. At the Rosatomnadzor NTC meeting on the prospects for the development and implementation of nuclear power projects in the Russian Federation (in Russian)
30. Ostretsov IN (2011) Nuclear energy now and in the future”, according to the site materials—newspaper “Zavtra” no. 12 (in Russian)
31. http://www.gosnadzor.ru—CFTP “Atomenergomash” (in Russian)
32. Kovalevich OM (2005) Safety of nuclear power plants. MEI, Moscow (in Russian)
33. Relex Faut Tree Module (FMEA)—473. Relex.0709. St. Petersburg, 2011 (in Russian)
34. Evdokimov VG, Komarova YV, Kuklev EA, Chinyuchin YM (2013) Quantitative estimation of the possibility of occurrence of accidents in aviation complexes. Scientific bulletin of the MSTU CA. no. 187(1). Moscow. pp 53–56 (in Russian)
35. Evdokimov VG. Methodical bases of safety assessment and risk management in aviation systems. International Aviation and Space Magazine “AviaSoyuz”. NQ 2 (35) April–May 2011 (in Russian)
36. Technical description of ACARS for A-380 aircraft. Express information, Moscow: 2010 (in Russian)
37. Kirpichev IG (2011) On the prospects and problems in developing the infrastructure for the maintenance of An-140, A-148 aircraft. Aircraft construction. Aviation Industry, no. 2. Moscow, 55 p (in Russian)
38. Gipich GN, Evdokimov VG, Chinyuchin YM (2013) Basic provisions of the concept of building a safety management system for aviation activities. Scientific bulletin of the MSTU CA. no. 187(1). Moscow. pp 31–36 (in Russian)
39. Accident Prevention Manual (1984) Doc. 9422-AN/923. International Civil Aviation Organization
40. Korolev VYu, Bening VE, Shorgin SY (2011) Mathematical foundations of the theory of risk. Fizmatlit, Moscow (in Russian)
41. Barzilovich EY, Kashtanov VA, Kovalenko IN (1971) On minimax criteria in reliability problems. ASUSSR Bulletin. Ser. “Technical Cybernetics, no. 3. Moscow, pp 87–98 (in Russian)
42. Smurov MY, Kuklev EA, Evdokimov VG, Gipich GN (2012) Safety of civil aircraft flights taking into account the risks of negative events. J Trans Russ Fed 1(38):54–58 (in Russian)
43. Kotik M.A., Emelyanov A.M. The nature of human operator errors (on examples of vehicle management). Moscow: Transport, 1993 (in Russian)
44. Kozlov VV (2008) Safety management. OJSC “Aeroflot”. Moscow (in Russian)
45. Bykov AA, Demin VF, Shevelyov YV (1989) Development of the basics of risk analysis and safety management. Collection of scientific papers of the Kurchatov Institute of Atomic Energy. Moscow: Publ. House IAE (in Russian)
46. Livanov VD, Novozhilov GV, Neymark MS (2013) Flight safety management system. IL SMS. “AviaSoyuz”, no. 1, pp. 14–21 (in Russian)

Chapter 6
Assessing Safety of Dual-Purpose Systems

The review materials are presented that contain a summary of the main positions of the RRS doctrine and the SST tools in connection with the assessment of the prospects for the development of SMSs (AA SMS) and corresponding GOST-R standards in civil aviation of the Russian Federation, taking into account ICAO recommendations on the basis of Annex 19 for dual-purpose ATSs.

One of the areas of application of such SMSs is the helicopter industry, where safety issues are considered very deeply, but mainly from the standpoint of the requirement of operational documents (AFMs, instructions for piloting helicopters and rules for performing aerial works) [1–5].

In the helicopter industry, there are a large number of documents regulating the design and development of helicopters as general aviation, but there are also special industry requirements in the form of "OSTs". At the same time, international standards for safety management are also used.

6.1 Recommendations of ICAO Amendment No. 101 Regarding the Requirements for the Development of SMSs (AA SMSs) for Industrial Production

The content of ICAO Amendment No. 101 should be considered as an international standard for AA SMSs in the field of ensuring the safety of industrial production, products, and articles in the transport sector.

The importance of this amendment for the RF civil aviation is that SMSs should be created and implemented in civil aviation and in a number of industries that produce dual-purpose equipment. The main recommendation of the amendment for the issue is the need to provide acceptable levels of safety for the ATS use throughout the life cycle of ensuring the functional worthiness and airworthiness of products and processes for aviation equipment operation, taking into account the design and

Kuklev E.A. et al., *Aviation System Risks and Safety*, Springer Aerospace Technology, https://doi.org/10.1007/978-981-13-8122-5_6

production stages. SMSs should be created taking into account this recommendation for both operators and service providers (according to Annex 19).

In this regard, the scientific provisions of the new doctrine "Reliability, risk, safety" are proposed to be used as a basis for a consistent (non-contradictory) combining of the RT requirements to the quality of industrial complexes and to the safety of the manufactured equipment on the basis of the risk indicators set out in the SST (Chaps. 1 and 2).

The problematic issues that need to be addressed are:

1. Definition of a list of standards required to ensure compliance of technical and economic characteristics (TECs) of products with Amendment No. 101.
2. Development of procedures for *recalculating the residual production risk* into ATS operational risks during their life cycle.
3. Standardization of modules and procedures for risk analysis and ATS state management taking into account hazards in the MCFs ("minimal cut set of failures") and in J. Reason chains.
4. Creation of the classifier of industrial safety by types of articles production.
5. Development of procedures for analysis of risk trends, depending on ATS key hazard factors from possible threats (IOSA type standard).

6.1.1 Classifier of Industrial Safety Types in the System Safety Theory

The need to create a classifier of this type, industrial safety (IS), is associated with a large difference in safety requirements for products and samples of equipment in different industries. This applies in particular to the specifics of the methodology for calculating the risks of adverse consequences during the life cycle (LC) of products and articles. The functional SST module is proposed as a basis of the IS classifier (from Chap. 2 of the book), as well as the principle of assessing the significance of the risks of rare events such as "catastrophes" with the "near-zero" probability. Thus, from the classical RT follows the need to take into account, according to I. Aronov (Chap. 1), the factors of "passive safety" ($F1$—in the form of design requirements usually controlled by industry acceptance procedures) and operational factors ($F2$) included in the rules for the operation of industrial products. This approach makes it *possible to monitor the safety state on the basis of the acceptable risk concept* [6–11].

The types of industrial safety (IS) are as follows.

(a) Types of industrial safety by factor F1, which are determined by standards for production of "cars", "air and sea vessels", "buildings and structures". Within the classical theory of product reliability, the following principle is usually used: "to ensure the capability of equipment ($F1$) to withstand external force loads that vary over time". This distinguishes "structural safety" in the group of functional failures by "design factors"—$F1$, according to I. Aronov ("passive safety") [12].

(b) The safety of the system operation is determined by the second factor $F2$ for functional failures (FOs) from the "operational" class in the form of consequences and harm arising from equipment failures and multiple events from MCFs during operation taking into account the impacts and the manifestation of human factors (HF).

The general classifier of IS types will be as follows:

(a) ***PB-1—for energy systems and production*** as a function of factors $F1$ *and* $F2$ characterizing nuclear power plants, hydropower plants, chemical plants, transport hubs such as large airports (Domodedovo, Khitrovo, etc.) and seaports (St. Petersburg, Nakhodka, and others).

b) ***PB-2—for production complexes and corporations*** producing technology-intensive small-scale products significant in terms of IS indicators: production of aircraft in the Russian Federation within the UAC system, Boeing Corporation, Airbus, and also in the production of railroad and motor transport, the production of nuclear vehicles, including nuclear-powered submarines (NPSs).

(c) ***PB-3—production of various products safe for use by the human population*** (household appliances, medicines and food products, etc.).

The book proposes to use convenient indicators of the level of safety in each of the IS types through the level of risk per the SST—not only "probabilistic" ones. Thus, it is proposed to adopt the RRS doctrine.

6.2 Methodological Basis for Implementing the Recommendations of Amendment No. 101 on the Basis of ILS Principles

6.2.1 IS Monitoring Subsystems

The classifier of industrial safety types developed taking into account the RRS principles makes it possible to designate and apply practically two most important *indicators of production quality* in the IS sphere, such as "safety" and "reliability", in the simplest way [13].

It can be assumed that it is this approach in the ILS system with the use of the MEL and MSG strategy that provides a high level of industrial safety for Boeing and Airbus aircraft operated in civil aviation of the Russian Federation.

Three main subsystems of the ILS system are shown in Figs. 6.1 and 6.2.

It is shown that the ILS-based "*safety monitoring*" *allows maintaining the "residual risk"* at an acceptable level and even reducing operational risks if the MSG and MEL strategies are adopted (as in "Airbus" company) [14].

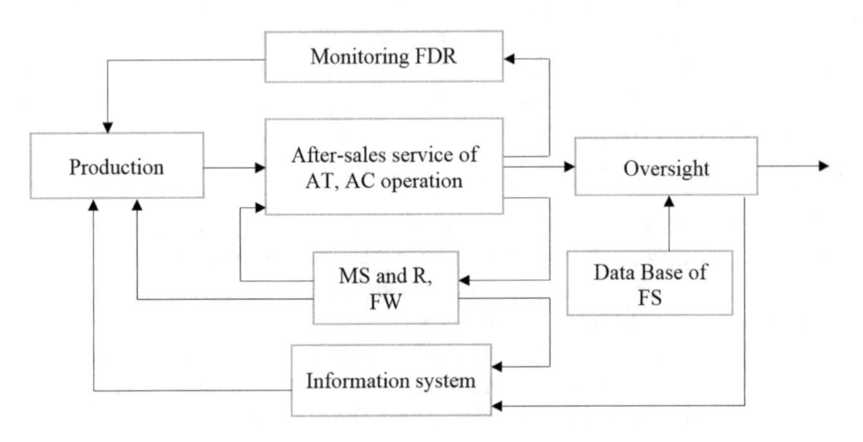

Fig. 6.1 QMS & SMS Interaction (upon A. Ynoussy)

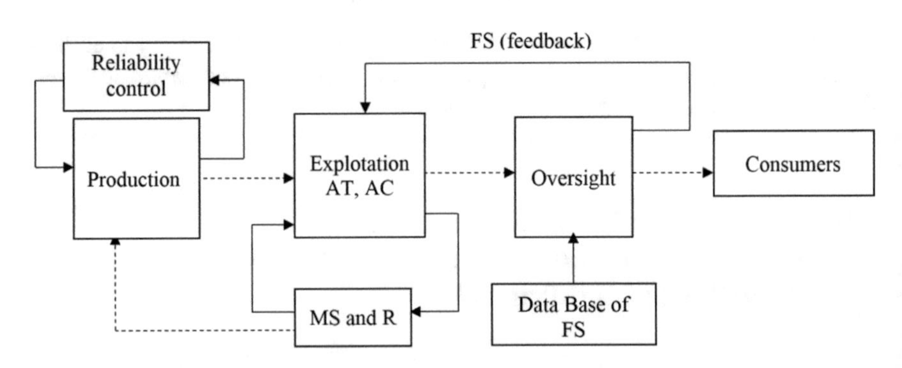

Fig. 6.2 Feedbacks in blocks for RS & FS

Figure 6.1 shows subsystems for maintaining aircraft airworthiness without feedback paths typical in the past for traditional production of aircraft and ATSs in the Russian Federation. This was done on the basis of the classical RT methods, i.e., without adjusting for flight safety indicators. Currently, the adjustment for flight safety is mandatory.

In this scheme, the after-sales service of AE is practically absent from the manufacturer's list of spare parts supplies. Proactive flight safety management on the part of the manufacturer by factors $F1$ *and* $F2$ is provided insufficiently.

Monitoring and control of *"residual risks"* for ATSs by $F1$ are not fully predicted.

In these schemes, the *after production "residual risk" can be significant*. However, with the transition to the *continued airworthiness technology by the "technical condition"*, taking into account the flight safety index, the MEL program, and the new standards, it is possible to keep the "operational risk" $\hat{R}_0$ at an acceptable level: Equation Chapter (Next) Sect. 6.6

$$\hat{R}_{0x} \sim \left(\hat{R}_0 < \hat{R}_{0x}\right). \tag{6.1}$$

A similar scheme was introduced into the K-32 helicopter operation system [15]. In particular, it is allowed to maintain the residual risk $\hat{R}_0$ in the industrial safety with the level of the risk event R by probability at the following level:

$$\mathrm{P}(R|\ \Sigma_0) \sim \mu_{1P} \cong 10^{-6}. \tag{6.2}$$

Other levels cannot be assigned, as it will be not provable and unacceptable. But worse values of up to ~10^{-3}, 10^{-4} are allowable, although undesirable. These are allowable, since due to new maintenance and repair technologies it is possible to keep the overall level at an acceptable level of risk.

6.2.2 Functions in the ILS System for Airbus Aircraft

An overview of the continued airworthiness functions of these aircraft is provided here in order to demonstrate the benefits of flight safety management methods based on risk calculation (e.g., per the RRS).

The features of the ILS system under consideration per in Fig. 6.2 are as follows.

1. The after-sales service of aircraft with the supply of spare parts from the manufacturer for maintenance and repair is provided throughout the life cycle of ATSs.
2. Mandatory monitoring of risks in civil aviation using FDR and ACARS for checking residual risks and reliability indicators provided by the ATS manufacturer.
3. The reliability of the aircraft functional systems is monitored by the acceptable level of risk $\tilde{R}_{0*}$, i.e., $\mu_1 \sim P_0$ based on redistribution of risk by the PSA method (by I. Aronov).
4. Proactive management of calculated risks in airlines (or by providers) is provided taking into account the results of the PSA, the safety assessment for aviation activities in accordance with the level of acceptable risk based on integral criteria with an assessment of financial costs for flight safety and compensation for non-pecuniary damage. Such technology of flight safety provision even with high *"residual risks"* allows reducing the cost of aircraft production, as *high reliability of product complexes due to multiple redundancy is not required*. A model of the ILS system application in Russian aircraft is the system of maintenance and repair planning, and maintaining continued airworthiness of AN-148 aircraft.

The Antonov Experimental Design Bureau (Ukraine) developed guidelines for the MEL and MMEL programs by analogy with Boeing aircraft.

In addition, the MMEL guidelines for AN-148 aircraft contain the rules for checking and *replacing units* on the basis of "risk indicators". The *risk assessment methodology is not presented*, but there are references to the ATS manufacturer [16].

The noted problem is solved quite simply within the RRS doctrine. At the same time, it is possible to explain the essence of the ATS manufacturer's procedures. But for this, it is necessary to create a set of corresponding new standards in civil aviation of the Russian Federation. The minimum number of various standards required to implement the above methodology for calculating risks is about 500 documents.

6.3 Evaluation of the Prospects for Transition of Civil Aviation of the Russian Federation to the New IS Standards and Provision of After-Sales Services for Industrial Production (F1 Factor) and Operation of Equipment (F2 Factor)

6.3.1 Status of Development

It was noted above that the systems of the type considered are developed and implemented at various scales of performance and significance in the Western aviation community in relation to the production and operation of Airbus and Boeing aircraft, types A-320, B-737, A-380, B-747, B-767, etc. The regulatory requirements for the system are set forth in the IOSA (IATA) documents [17–19]. The main characteristics of the ILS systems for the production and operation of aircraft are as follows: maintenance of standard safety indicators, organization of the maintenance and repair system in accordance with the principle of continued airworthiness of aircraft according to their technical condition, flight safety management through SMSs, analyzing aircraft flight parameters using onboard recorders FDRs, and real-time monitoring of aircraft flight parameters based on ACAR (ACARS) systems for modern aircraft similar to A-380 [14].

Possible areas of IS system modernization are identified here as first approximation on the basis of analysis of the results and consequences of known *"accidents"*, *"catastrophes"*, and abnormal *"natural phenomena"* (such as *"floods"*, *"tornadoes"*, *"snowfalls"*, *"temperature drops"*, *"massive fires"*).The main areas are the following: *abandoning the principle "if it is reliable, then it is safe"* and the transition to the principles of "risk calculation" according to Annex 19, including RRS and SST principles.

6.3.2 *Structure of the Set of Standards*

The Ministry of Transport of the Russian Federation acknowledged that it is necessary to create an SMS of international type in the civil aviation of Russia in the form of AA SMS based on ICAO SARPs and Annex 19 [8].

The main requirements and the list of SMS modules are presented in the paper of Amir Yunossi's group from the FAA (in the "Blue folder" [20] as described in Chap. 4 of this book).

The method of probabilistic safety analysis is described in the NASA document (2002–2007) "Probabilistic Risk Assessment Procedures for NASA Managers and Practitioners" (Version 2.2.—Washington, DC 20546, Aug. 2002)—its content is presented in the appendix to the book.

The structure of the set of standards required to regulate aviation safety regarding helicopters is given in Fig. 6.5. The issue of creating standards for helicopters (as GPA) is considered to be a priority, as noted above, since the number of such documents in this domain is small. On the scale of civil aviation of the Russian Federation, the number of standards in the field of flight safety is also small, but at least international trends in Annex 19 are taken into account.

6.4 MSG Strategy for the Development of a Maintenance and Repair Program (Reliability) for Western-Made Aircraft

Some features of these programs are considered in order to assess the possibility of increasing flight safety based on the recommendations of Annex 19 and the RRS doctrine.

6.4.1 *Maintenance Program Structure*

The baseline of the programs is the requirements for ensuring flight safety, reliability (fail-safety), and maintainability (serviceability).

1. If a failure or combination of failures of elements affects flight safety, then with increasing failure rates, the following options are possible: The element is serviced according to the operating time or considering its state with some parameter control. However, the documents [21] do not address the following issues: How to assess the change in safety levels for various maintenance and repair strategies?

6.4.2 Aircraft Maintenance and Reliability Assurance Programs in MSG-1, MSG-3

Such programs are mainly studied and developed in FSUE State Research Institute of Civil Aviation (GosNII GA).

Three phases are established for the production and application of ATSs with monitoring of parameters in scenarios with functional failures.

Stage* 1 *(Phase* 1*)—achievement of standard RT indicators on the basis of the PF without taking into account the requirements for industrial safety.

Stage* 2 *(Phase* 2*)—providing indicators of high reliability of systems, taking into account the requirements of industrial safety and norms of the “residual risk” during their life cycle, based on the industrial ILS strategies for Ka-32 helicopters (research of “Aviatekhpriemka”, research director is Evdokimov V.G.).

Stage* 3 *(Phase* 3*)—assessing and maintaining the level of system safety (IS) by managing risk parameters based on proactively identified risk factors and types of risks for selected systems by functional failures.

The action strategy in civil aviation of the Russian Federation within the RRS doctrine in the transition to the IATA ILS system

The main requirements are as follows: The *system produced must be highly reliable*; the manufacturer of the equipment must provide quality (RT property) such that the “residual risk” ***by the probability of a risk event*** is ***not worse than*** $\mathbf{10^{-4}}$–$\mathbf{10^{-6}}$. (This indicator is established by the ATS developer in agreement with the operator).

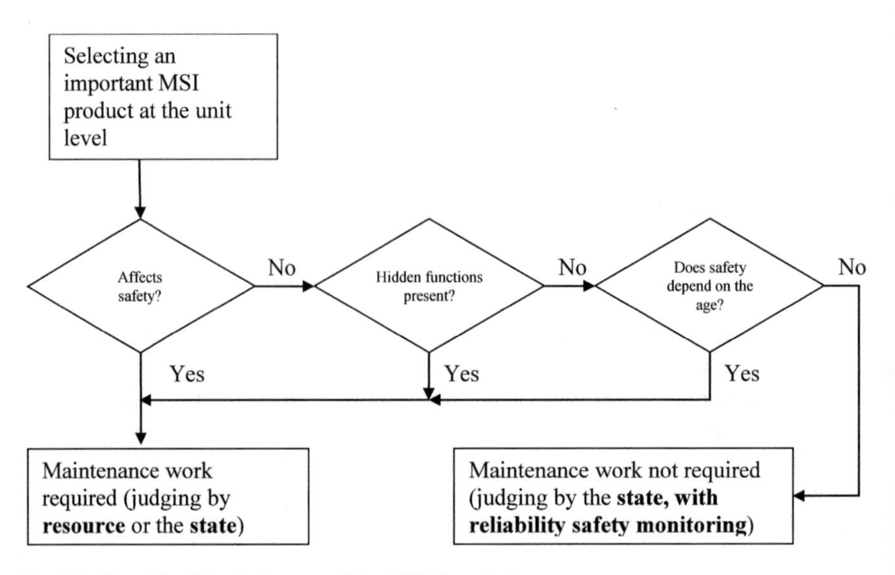

Fig. 6.3 Simplified block diagram of the MSG-3 analysis

MSG-1/2	HT	OC	CM	–		–
	MSG-3	DS/RS	IN/OP/FC	–	SV	LU
Affects safety	With the MSG analysis	X	X			
Hidden functions present		X	X			
Affects efficiency		X	X		X	X
Affects operation		X	X		X	X
Authority requirements		X	X			
Limitation of service life		X				

Fig. 6.4 Selection of maintenance methods

CONTROL DIAGRAM
FOR HELICOPTER FLIGT-SAFETY
SERVICE PROCEDURES

A. Modules of "Flight worthiness"

A1 *– Logistic and lysis of the Maintenance*

A2 *– Development of Maintenance and Repair System*

A3 *– Development of Suporting System for Flight Worthness*

B. Contents if A – procedures

A1 *– Reliability.*

A2 *– Concept of MRS/ Development*

A3 *– Codifications. Initial Parameters of MRS. Control system of Details Supply and Scores for Bying of Helicopters.*

Notes: **MRS** – Maintenance & Repair System

Fig. 6.5 Scheme of FS-service procedures

In developing a system as a whole and AA SMS modules for JSC "Russian Helicopters", provisions from the AA safety management methodology and principles of risk calculation and management are applied taking into account the recommendations of the ICAO Annex 19 (version of 2012).

When operating systems, the second factor $F2$ is considered—an "operational" one characterizing effects and harm to consumers due to "failures" of products or ATS system as a whole when using the AE influenced by the external environment and existence of internal adverse factors in the system (with loss of AE functional properties) (Fig. 6.3).

The scheme for analyzing ATS properties according to MSG-2 [16] is shown in Figs. 6.4 and 6.5.

6.5 Design Requirements for Ensuring Flight Safety of Helicopters with an External Cargo Sling Load System

6.5.1 Methodical Approach to the Formation of the Logistic Support System for the After-Sales Service of Ka-32 Helicopters

A helicopter of this type is a sample of dual-purpose aircraft. In this regard, this section provides a scheme for implementing the methodology of logistic support for ensuring the airworthiness of dual-purpose aircraft using the example of the Ka-32 helicopter [1].

The task of creating an SMS within a set of requirements for ensuring industrial safety is solved, taking into account the recommendations of ICAO amendment No. 101 on how to compensate for the residual risk due to systematic errors in the design and manufacture of Ka-32 helicopters. For helicopters such as MI-27, K-32, design features of the surveillance system in the cargo cabin of the external sling load system should be known, especially when cargo is transported by two helicopters with some beam suspension system.

It is important to establish requirements for ensuring flight safety in the performance of aerial works with cargo on external suspension.

The basic theoretical provisions adopted in this book are that when developing an SMS, the ICAO definition is taken into account: "*Risk is an amount of hazard in the system by the factors of random occurrence of a risk event and damage*", which is constructive. Types, consequences, and criticality of failures of the main units, components, and assemblies of the system safety are analyzed according to Technological Diagram [8]:

It is proposed to consider some features of the development of the classification of industrial safety types in an AA SMS for helicopters.

6.5.2 Recommendations for Helicopter SMS Development Strategy

The AA SMS for helicopters should be developed with the requirements of ICAO Amendment No. 101 in mind. Therefore, it is necessary to take into account the features of the industrial safety concept and not only the type of flight safety that is studied in operation of various types of aircraft. The key point in this case is the identification of factors of "structural safety", which significantly affects the choice and maintenance of an ***acceptable level of residual risk***.

In the case of helicopters, these are factors ***F*-1** (in *IS*-2), for example, onboard display facilities, navigation instruments, communication with ***GLONASS or GPS*** satellite systems, and the location of the cargo suspension assemblies (at the center of mass or at the bottom of the hull, presence of sensors detecting the resonance of helicopter blades, vibrations of engine shafts and bearings, etc.). To this end, the following known principles, as formulated above (by ICAO), are adopted for the SMS.

Principle No. **1**. Ensuring high reliability of the system, in particular, by creating and applying multiple protection lines. According to the PSA recommendations (from Chap. 1 of this book), it is shown that the guaranteed residual risk $\Delta\hat{R}$ for the system in terms of the probability of a risk event should be typical at the level of 10^{-4}–10^{-6}—not worse—per year or for a given period T of system operation (i.e., up to 10^{-4}–10^{-6} for the frequency of one catastrophe per year, e.g., for NPPs, according to IAEA). Then the integral risk $\Delta\tilde{R}$ must correspond to the PSA norms for the risk event R:

$$\Delta\hat{R} \sim (R/\Sigma_0)\ 10^{-4} - 10^{-6}(\text{for the NPP life cycle})$$

Classification of ATSs and types of AA as hazardous and safe when using helicopters for aerial works is performed on the basis of the ICAO risk concept by proactively establishing the possibility of occurrence of hazardous (risk) events with two properties in ATSs.

Principle No. **2**. Assessing risks of catastrophes based on the provisions of the new doctrine (RRS from the SST) by searching for catastrophes using chains of events such as J. Reason chains without using probabilistic indicators, but with risk indicators and comparing them with acceptable levels of risk.

Principle No. **3**. Synthesis of preventive corrective influences on the system for ensuring safety taking into account the risk factors on the set of identified possible paths to a catastrophe (without the PSA methods and *without probabilistic indicators*) by the ICAO tuple (4.8) from Chap. 4. The structure of the safety assurance system and the characteristics and the application of a given number of protections are changed considering the system safety indicators if these protections provide a method for reducing risks of catastrophes.

6.6 Importance of the New RSS Ideology (Adopted in the SST for Flight Safety Evaluation) for Science and Practice in Comparison with Russian and Foreign Approaches to the Construction of Safety Management Systems Based on the Calculation of Risks

6.6.1 Assessment of the Significance of RRS Methods for Evaluation of ATS Operation Safety

In light of the new RRS doctrine, one has to consider the PSA approach intensively developed in the classical RT in the last two decades, as conditioned by the rigid need to solve the problem of evaluating ATS safety level for rare events, but *within the principles of the hypercube of truth (Boolean lattice)* [15] that proved to be difficult.

The strongest positions in the PSA method that were put forth within the classical RT by Malinetskiy [22] ("heavy tails") and M. Fujita *"On confidence domains for determining the values of the probabilities of processes in the regions of pdf "tails"* [23] cannot solve the problem.

Malinetskiy [22] *proposed, in fact, a constructive, very important way of searching for the lower limits of pdf "tails"*, but also, as *M. Fujita, could not find a way to determine the reliable formula for pdf.*

The actual tasks are: *confirmation of the priority importance of the RT to ensure high reliability of systems* up to the boundaries of the still distinct significance of the probability of a risk event—up to 10^{-6} (not better, according to the GOST-R standard). The second aspect of this statement boils down to the fact that it is necessary to *ensure, with the help of the PSA, the solution of only correct tasks to assess the operability of systems* using the PF of LAFs type and others and at the same time ***transfer the solution of all "safety" issues to the domain of fuzzy subsets***;

In connection with this, it is proposed to substantiate in the *SST a completely different apparatus of a "fuzzy subset method" type to find solutions for assessing system safety based on the logic of calculating risks in fuzzy measures.*

6.6.2 List of Projects of Scientific and Technical Research on the Implementation of the SST Provisions in Flight Safety Management Systems

Themes of possible projects

The development of standards for the calculation of risks and the substantiation of SMS requirements based on the SST provisions should be recognized as the most important research domain. At the same time, it is necessary to take into account aspects of the "rare events problem", methods for constructing J. Reason chains, hazard models (by ICAO, per the SST), to consider physical and Boolean bases of the universal (clear) set of facilities of real technical systems, to apply the "minimal cut sets of failures", etc.

Domain No. 1. *Standardization of SMS terms, definitions and structures, taking into account the provisions of the RRS doctrine together with the SST.*

At the Assembly No. 37 (October 2010), ICAO adopted a resolution to establish a working group with representatives from the IAC, the RF, and the Air Navigation Commission (ICAO) to develop issues in this domain in connection with the preparation of the ICAO standard for SMSs (in the form of Annex 19).

Domain No. 2. *Organization of cooperation with "Airbus" and "Boeing" corporations (within reasonable limits).*

This is necessary, since the practical achievements of the aviation community in the application of the methodology for calculating risks for assessing the flight safety level are quite significant.

This is especially noticeable in the development of technologies to maintain the airworthiness level of operated aircraft, taking into account the *maintenance and repair* systems based on the *MSG and MEL* strategies.

The content of procedures and algorithms to ensure industrial safety includes the development of methods to maintain functional worthiness (or functional reliability) in the field of production and operation of dual-purpose equipment with the provision of standard reliability indicators in the life cycle of products.

The *scientific "breakthrough"* is the development of theoretical foundations for solving the rare events problem based on the application of the methodology for assessing the quality of functioning of a complex system with *fuzzy subsets* of hazardous (risk) event classes predetermined in the classical RT.

The *technical result* should be considered as the practical application of the developed approach on the examples of SMSs in civil aviation in the form of algorithms, procedures, and computer programs providing creation of databases on risk factors, identifying conditions for the occurrence of catastrophes and the development of safety management methods in civil aviation by changing current and predicted states of systems taking into account manifestation of risk factors in systems. The effectiveness of applying new approaches such as "J. Reason chains", as well as the method of creating barriers and proactive management of the system state based on "*common sense*", as in Arzamas-16, is established.

The list of indicators of the AA safety level regulated by ICAO through Annex 19 is defined within the methodology for calculating risks using special tools for measuring and identifying risks and threats based on approaches to solving the rare events problem (per ICAO—events with the "near-zero" probability) without the use of probabilistic indicators and PSA calculations.

The methods of the classical reliability theory are used as tools for creating highly reliable systems at the stages of design and manufacturing helicopters and maintaining airworthiness at various stages of the products life cycle.

6.7 Conclusions

1. The main result of this chapter is that when assessing risks in the field of fuzzy subsets, it becomes correct and simple to calculate the risks only for "damages" and only for one value of the event randomness measure with the "near-zero" probability. This is justified by the fact that the level of high reliability of the system and the "rarity" of hazardous events is guaranteed through the introduction of quality management systems and "reliability" standards, especially for dual-purpose equipment.
2. *The task is to move to the new ICAO flight safety (and industrial safety) programs with the new doctrine "Reliability, Risk, Safety" within the system safety theory developed with due regard to Annex* 19.

References

1. Evdokimov VG (2011) Integrated logistics support for the production of K-32 helicopters.—"Aviatekhpriemka", Moscow. (in Russian)
2. CAP 426 (2006) Helicopter external load operations/safety regulation group. West Sussex 2006
3. Instructions for transportation of goods on the external suspension of helicopters. MTR of the Russian Federation (№ Kr-2-r of 8.01.2004). (in Russian)
4. Safety provisions (2010) Helicopter European safety group and Jsc Allied Aviation Consulting—Ehest (ESSI). JSC "AAC", Moscow (in Russian)
5. Guidance on hazard identification (2010) SMS WG (ECAST (ESSI) Working group on safety management system and safety culture). Handwritten. (in Russian)
6. Standard for the assessment of major aviation risks. Version 4 (Threats). (Raw material industries). Flight Safety Foundation. Melbourne (Australia), 2012. (in Russian)
7. Gipich GN, Evdokimov VG, Kuklev EA (2010) Methodical provisions of the classifier "SMS terms and definitions" in the field of flight safety management. Coll.of papers of the NTC "Rostekhregulirovanie", vol. 2. Moscow. (in Russian)
8. Gipich GN, Evdokimov VG (2011) The concept in the field of standardization of principles and procedures for the development of a national version of the system and manual for safety management in the production of aircraft in the UAC at the state level in the Russian Federation. Report (printed) in the "UAC Bulletin" vol. 2 1. Moscow, pp 65–70 (in press). (in Russian)
9. Evdokimov VG (2009) Scientific basis for the implementation of the state targeted program for the safety of aviation equipment in civil aviation of the russian federation in 2010–2011 based on the requirements of integrated logistics support for production. J Air Trans. Moscow, pp 32–37. (in Russian)
10. Evdokimov VG, Gipich GN (2006) Some issues in the methodology of assessment and prediction of air transport risks. In: The 5th international conference "Aviation and Astronautic Science 2006", Abstracts, Moscow pp 75–76. (in Russian)
11. Evdokimov VG, Grushchanskiy VA, Kavtaradze EA (2008) On the system safety of information support for the development of complex social systems. Fundamental problems of system safety: Coll. of papers/RAS Dorodnitsyn CC. Moscow, University Book, pp 67–77. (in Russian)
12. Aronov IZ et al (2009) Reliability and safety of technical systems. Moscow. (in Russian)
13. Gipich GN, Evdokimov VG, Shapkin VS (2013) Assessments of the system safety for industrial technical complexes on the basis of the theory of risks in the aviation industry and in the field of nuclear energy. Scientific bulletin of the MSTU CA. No. 187(1). Moscow, pp 46–49. (in Russian)

14. Technical description of ACARS for A-380 aircraft. Express information, Moscow, 2010 (in Russian)
15. Ryabinin IA (1997) Reliability, survivability and safety of ship electric power systems. Kuznetsov Naval Academy, St. Petersburg. (in Russian)
16. Kirpichev IG (2011) On the prospects and problems in developing the infrastructure for the maintenance of An-140, A-148 aircraft. Aircraft construction "Aviation Industry", vol 2. Moscow , 55 (in Russian)
17. Risk management—Vocabulary—Guidelines for use in standarts. PD ISO/IEC, Guide 73: 2002. B51: 2009
18. Evdokimov VG (2010) Harmonization of Russia's regulatory and legal framework in the field of industrial safety. Coll. of papers "Aviatechobozrenie", vol. 8. Moscow, pp 83–89 (in Russian)
19. Evdokimov VG (2011) Acceptable level of safety of the Russian aviation system. International Aviation and Space Magazine "AviaSouz". NQ 1 (34) January–February (in Russian)
20. Amer MY (2012) 10 things you should know about safety management systems (SMS). SM ICG, Washington
21. Gosatomnadzor of the Russian Federation (1997) General provisions for ensuring the safety of nuclear power plants. PNAE G-I-011-97 (in Russian)
22. Malinetskiy GG, Kulba VV, Kosyachenko SA, Shnirman MG et al (2000) Risk management. Risk. Sustainable development. Synergetics. Moscow, Nauka, Series "Cybernetics", RAS. p 431 (in Russian)
23. Fujita M (2009) Frequency of rare event occurrences (ICAO collision risk model for Separation minima). RVSM. ICAO, Doc. 2458. Tokio: EIWAC 2009
24. SMM (Safety Management Manual): Doc 9859_AN474–Doc FAA: 2012
25. General rules for risk assessment and management (resource management at life cycle stages, risk and reliability analysis management—URRAN). JSC Russian Railways standard STO RR 1.02.034-2010. Moscow, (2010) (in Russian)

Conclusion

The creation of the RRS doctrine was the result of the scientific community reaction, while recognizing the importance of "common sense" in matters of safety, to external challenges in the aviation industry. The perfection and beauty of the mathematical apparatus in the form of RT, PSA, and simulation modeling cannot provide the necessary reliability of the solution of the "rare events" problem declared by ICAO as a priority. "The Western world" with its powerful aviation industry and the need of the nations all over the world for transportation made every state recognize their technologies based on the calculation of risks of negative consequences and proactive methods of safety management. To do this, it is enough to note one of the effective programs called MEL or MMEL. It is obvious, for example, that no "Markov chains" can properly describe possible random changes in ATS states on the basis of transition matrices for which there is no statistics on the transition probability.

In the practice of operating foreign-made military aircraft, small values of the probabilities of events in a number of processes observed during the ATS operation are completely non-informative, and therefore, a different approach should be proposed.

The "secrets" of risk assessment and databases are "deeply hidden" in the centers of Toulouse and Seattle, and the results of their research are difficult to access for civil aviation of the Russian Federation.

The essence of the difficulties is that "ATS safety", according to the classical RT, should be determined on the events "opposite to the main event" in the same binary outcome space—opposite to the "operability" event (consumer quality, according to A. Younossi from USA FAA) [1, 2].

By tradition, any random event must have a non-random measure of measurability, for example, in the form of "probability". "How to work without probabilities in this random world?"—that is the most challenging issue of aviation practitioners. But rare events, for example, the catastrophe with the "Concordia" (from Chap. 5 of this book) cannot be strictly attributed to random ones, since there are no reliable pdf and Prdf for them, as they all lie in the region of "heavy tails" ("…tail-far from

Kuklev E.A. et al., *Aviation System Risks and Safety*, Springer Aerospace Technology, https://doi.org/10.1007/978-981-13-8122-5

medium …", according to M. Fujita). A measure of the possibility of the emergence of ATS states that characterize the properties of "rare events" does not correspond to the "truth" provisions on the Boolean lattice from the RT. It was possible to establish in the book that the basis should be fuzzy models of ATS processes within, for example, Fuzzy Sets. This domain of research is quite promising, what excuses authors to some extent and therefore is discussed in this book.

The materials of the book are arranged in regard to the need to demonstrate the usefulness of the SST tools in solving a number of difficult problems in the field of flight safety and aviation activities. In particular, the decision-making methods based on the weighing of "risks" and "chances" were examined without probabilistic measures in projects concerning ensuring ATS efficiency.

One of the main issues is the meaning and significance of J. Reason chains for the processes of flight safety (system safety) proactive management using standard SMSs. In fact, there is nothing new in J. Reason chains. These are just event scenarios, similar to scenarios from the event tree method in the FMEA program (in the RT). However, these chains proved to be universal in terms of construction and more understandable for their application in solving complex problems. In this book, in view of this, methods are given for constructing such chains using combinatorial methods and evaluating the criticality of the analysis results regarding the ATS safety state without probabilistic indicators.

It is proved that the *methodology for the correct calculation of risks* can be completely and reliably developed only in the field of assessing flight safety in civil aviation with the application of procedures based on the *Fuzzy Sets doctrine* to assess the criticality of event chains or scenarios without probabilistic indicators.

It has been revealed that managing risks of accidents and *improving system safety* using quality assurance methods only by *fail-safety* of equipment is not sufficiently correct.

The most important result following from the FAA document under consideration for the Russian aviation industry is the recommendation that SMSs can be integrated into the existing quality systems at each aviation enterprise. (As is known, this idea was recognized in IATA three years ago in the form of an integrated quality system, as reported in 2010 on behalf of TK-034 at the GosNII GA NTC meeting). The proposed concept can serve as a basis for determining the content of a number of projects on "system safety" including the project "Stabilizing the state of industrial safety" (SSIS) considered here.

The IS and SST theories based on the new risk calculation doctrine "Reliability, Risks, Safety" focus on the border of the provisions of the classical reliability theory; they may become common for all types of safety, including transport safety.

For example, in civil aviation and JSC Russian Railways (JSC RR), there are divergences in this area. In [3], the ISO definition "Risk is a probability … of an event …" is adopted, but the indicator "Risk probability …" is further proposed, which is inconsistent with the first concept. In civil aviation more proper (correct) ICAO (Annex 19) definitions and terms are used.

It is necessary to create a methodology for interpreting the concepts of *"challenge", "threat", "hazard", "safety", "damage", "acceptable risk",* "residual risk" in order to develop principles and methodologies for a uniform approach for assessing various types of safety in various activities of Russian enterprises, including financial, political, nuclear, food, transport, aviation safety, etc.

References

1. Amer MY (2012) 10 things you should know about safety management systems (SMS). SM ICG, Washington: 2012
2. SMM (Safety Management Manual): Doc 9859_AN474–Doc FAA: 2012
3. General rules for risk assessment and management (resource management at life cycle stages, risk and reliability analysis management—URRAN). JSC Russian Railways standard STO RR 1.02.034-2010. Moscow: 2010 (in Russian)

Printed in the United States
By Bookmasters